Selected Topics on Improved Oil Recovery

Berihun Mamo Negash · Sonny Irawan
Taufan Marhaendrajana
Hasian P. Septoratno Siregar · Sudjati Rachmat
Luky Hendraningrat · Andi Setyo Wibowo
Editors

Selected Topics on Improved Oil Recovery

Transactions of the International Conference
on Improved Oil Recovery, 2017

 Springer

Editors

Berihun Mamo Negash
Department of Petroleum Engineering
Universiti Teknologi PETRONAS
Seri Iskandar, Perak
Malaysia

Sonny Irawan
Department of Petroleum Engineering
Universiti Teknologi PETRONAS
Seri Iskandar, Perak
Malaysia

Taufan Marhaendrajana
Department of Petroleum Engineering
Institut Teknologi Bandung
Bandung, Jawa Barat
Indonesia

Hasian P. Septoratno Siregar
Department of Petroleum Engineering
Institut Teknologi Bandung
Bandung, Jawa Barat
Indonesia

Sudjati Rachmat
Department of Petroleum Engineering
Institut Teknologi Bandung
Bandung, Jawa Barat
Indonesia

Luky Hendraningrat
PT. Auroris Oil and Gas
Jakarta
Indonesia

Andi Setyo Wibowo
LEMIGAS Production Engineering Group
Jakarta
Indonesia

ISBN 978-981-10-8449-2   ISBN 978-981-10-8450-8   (eBook)
https://doi.org/10.1007/978-981-10-8450-8

Library of Congress Control Number: 2018932186

Printed on acid-free paper

This Springer imprint is published by the registered company Springer Nature Singapore Pte Ltd. part of Springer Nature
The registered company address is: 152 Beach Road, #21-01/04 Gateway East, Singapore 189721, Singapore

Preface

This book is the outcome of the first Joint International Conference on Improved Oil Recovery (JICIOR 2017) between Petroleum Engineering Department of Institut Teknologi Bandung (ITB), Indonesia, and Petroleum Engineering Department of Universiti Teknologi PETRONAS (UTP), Malaysia. The conference is hosted by ITB, Indonesia. The chapters in this proceeding are peer-reviewed. The theme of the conference "Improved Oil Recovery" was aimed to attract papers on research that addresses the oil and gas industry with out-of-the-box approaches and pushing the boundaries of conventional applications on oil and gas production. The theme also implies the need to consider current and future technologies in order to meet oil and gas demand as well as accommodate with the challenge of low oil and gas prices. The subjects range from experiments and mathematical modeling works. We hope that our readers will find this proceedings beneficial, and perhaps we will meet at Joint International Conference 2018 hosted by UTP. The program schedule and all information regarding the conference may be accessed from the home page http://tm.itb.ac.id/en/2017/11/17/joint-conference-improved-oil-recovery/.

Seri Iskandar, Malaysia	Berihun Mamo Negash
Seri Iskandar, Malaysia	Sonny Irawan
Bandung, Indonesia	Taufan Marhaendrajana
Bandung, Indonesia	Hasian P. Septoratno Siregar
Bandung, Indonesia	Sudjati Rachmat
Jakarta, Indonesia	Luky Hendraningrat
Jakarta, Indonesia	Andi Setyo Wibowo

Acknowledgements

We would like to express our gratitude to all who have submitted and presented their research work at JICIOR 2017 which took place at West Hall (Aula Barat) of Institut Teknologi Bandung (Bandung, Indonesia) on November 30, 2017. We would also like to appreciate external and internal reviewers for the tremendous efforts and dedication of their valuable resource and time while reviewing 19 papers submitted. Moreover, we would like to extend our utmost appreciation to the four esteemed plenary speakers for delivering their respective fascinating talks in line with the theme of the conference "Improved Oil Recovery". Professor Dr. Ir. Doddy Abdassah, M.Sc., from Petroleum Engineering Department of Institut Teknologi Bandung has spoken about the ability of future oil and gas production in Indonesia in terms of Natuna field resources. Professor Eswaran Padmanaban from Universiti Teknologi PETRONAS discussed the terminology and future research on Improved Oil Recovery. Mohammad Rivai Lasahido, M.Sc., from SKK Migas presented the technology and resources in Indonesia from government point of view. Putu Suarsana, Ph.D., from Pertamina EP discussed Pertamina involvement on Improved Oil Recovery. We would like to thank our sponsors for the program from Medco Energy Persada and Pertamina EP.

Plenary Speakers

Prof. Dr. Ir. Doddy Abdassah, M.Sc., from Petroleum Engineering Department, Institut Teknologi Bandung

Prof. Eswaran Padmanaban from Universiti Teknologi PETRONAS

Mohammad Rivai Lasahido, M.Sc., from SKK Migas

Putu Suarsana, Ph.D., from Pertamina EP

Reviewers

Prof. Dr. Ir. Hasian P. Septoratno Siregar
Assoc. Prof. Dr. Sonny Irawan
Prof. Dr. Ir. Sudjati Rachmat
Dr. Andy Setyo Wibowo
Dr. Luky Hendraningrat
Dr. Taufan Marhaendrajana
Dr. Silvya Dewi Rahmawati
Rani Kurnia, MT

Conference Organizing Committee

Chairman
Dr. Taufan Marhaendrajana (ITB)

Co-chairman
Dr. Sonny Irawan (UTP)

Secretary
Dr. Silvya Dewi Rahmawati (ITB)

Treasurer
Firda Izzatturrohmah

Publication
Berihun Mamo Negash (UTP)
Rani Kurnia MT

Technical
Kharisma Idea MT

Sponsorship
Dr. Amega Yasutra

IT/Multimedia
Dr. Dedi Irawan

Webmaster
Widi

Logistic
Oman Rohman

Food and Beverage
Tuti Suhaemi

Event Management
Dr. Zuher Syihab

Promotion
Dr. Ardhi Lumban Gaol
Dr. Syahrir Ridha (UTP)

Public Relation
Dr. Silvya Dewi Rahmawati
Dr. Amega Yasutra

Secretariat/Registration
Firda Izzaturrohmah

Contents

Determination of Cementation Factor from Induced Polarization Concept

Wan Zairani Wan Bakar, Ismail Mohd Saaid and Suriatie Mat Yusuf

Abstract Determination of water saturation, S_w in clean sand is more or less straight-forward using the Archie equation. In shaly formation where the resistivity data is affected by clay conductivity, the Archie equation is no longer valid and S_w has to be modelled using other equations. The Waxman-Smits was one of the equations that has been developed to account for the clay effect in shaly sand. Nevertheless parameters used in the Waxman-Smits equation (i.e. formation resistivity factor, F^* and cementation factor, m^*) rely upon the expensive core analysis data, which is not necessarily available for the particular reservoir of interest. The current method in core analysis could also contribute to inaccurate value of the determined parameters; the issues on averaging and representation of the selected core plugs to the whole reservoir and the effect of core treatment to the rock properties. In this paper, we reviewed some previous works to understand mechanism of clay surface conductivity and induced polarization (IP) concept and proposed a potential method for determination of more accurate m^* using this concept.

Keywords Cementation factor · Formation factor · Clay conductivity
Complex conductivity · Induced polarization

Introduction

The Archie equation (Eq. 1) was introduced by Archie (1942) and had been used for decades to quantify water saturation, S_w. This equation was claimed to provide

W. Z. W. Bakar (✉) · I. M. Saaid
Department of Petroleum Engineering, Faculty of Geosciences & Petroleum Engineering, Universiti Teknologi PETRONAS, Seri Iskandar, Malaysia
e-mail: zairani@salam.uitm.edu.my

W. Z. W. Bakar · S. M. Yusuf
Department of Oil & Gas, Faculty of Chemical Engineering, Universiti Teknologi MARA, Shah Alam, Malaysia

© Springer Nature Singapore Pte Ltd. 2018
B. M. Negash et al. (eds.), *Selected Topics on Improved Oil Recovery*,
https://doi.org/10.1007/978-981-10-8450-8_1

good estimations of S_w in clean formation based on the resistivity log data that gives different response in water and hydrocarbon zones.

$$S_w^n = \frac{a R_w}{\phi^m R_t}$$ (1)

This equation was derived empirically based on a set of clay-free rocks obtained from the Nacatoch sandstone. The Archie type of rocks is described as water-wet formation comprises of nonconductive, approximately equidimensional grains with a simple unimodal intergranular pore system (Kennedy and Herrick 2012). Electrical conduction in this formation is through the electrolyte in pore spaces only. In this case saturated rock resistivity, R_o has a linear relationship with its saturating electrolyte resistivity, R_w and the produced constant is called the formation resistivity factor, F. This relationship is however does not applied to rocks with clay and conductive materials, oil wet reservoir and carbonates.

$$F = \sigma_w/\sigma_o = R_o/R_w$$ (2)

Archie reported that the formation resistivity factor, F is governed by the formation type and characteristics. This indicates that path of electrical current flow in an Archie's rock is controlled by size, dimension, arrangement and sorting of rock grains. Owing to this, formation factor has a power law relationship with the porosity;

$$F = 1/\emptyset^m$$ (3)
$$R_o = R_w \emptyset^{(-m)}$$ (4)

The physics of conductivity in Archie's rock was explained in a series of report by Herrick and Kennedy (1993, 1994, 2009) and Kennedy and Herrick (2012) who also proposed a geometrical factor theory. The geometrical theory provides an alternative to interpret conductivity of Archie's rock based on physical interpretation. In this theory, brine conductivity, fractional brine volume and the brine geometrical factor related to the spatial distribution were claimed to influence the bulk conductivity of Archie's rocks. The attempt to describe relationship between conductivity and pore geometry was previously done by Owen (1952). In his model a constriction factor was added, which was related to the ratio between pore throat length and its cross section exhibiting limitation of current flow that gives higher formation factors.

The Shaly Sand and Waxman-Smits Equation.

For Eq. (2) to be valid, formation resistivity factor, F of an Archie-type rock is expected to be constant over changes in brine resistivity, R_w. Nevertheless, in some types of rocks different observation was made whereby F was not a constant but decreased with an increased in R_w (Patnode and Wyllie 1950). This observation was particularly made in shaly sand, which suggested that clay minerals had introduced an excess conductivity at high R_w or low brine conductivity, σ_w (Worthington 1985). The bulk conductivity of the rock, R_o does not solely depend on the brine conductivity, R_w and the increase in R_o is not in linear proportionality to the increase in R_w.

In this case the linear equations proposed by Archie is no longer valid to be used to determine F or water saturation, S_w.

This phenomenon was subsequently studied by many researchers who had suggested equations to estimate water saturation in shaly sand accounting for this clay minerals conductivity. The development of shaly sand concept and equations used in the interpretation is discussed in detail in the paper by Worthington (1985). Most of the proposed equations attempted to quantify clay conductivity based on the volume of shale, Vsh. However, this assumption could be misleading since clay conductivity is not just a function of volume but also the type and distribution pattern (de Lima and Sharma 1990; Wildenschild et al. 2000).

Waxman and Smits (1968) introduced a model of two parallel conductance to describe the shaly sand conductivity (Eq. 5). The conductance elements in this model consist of brine conductance, C_w and clay conductance, C_{clay}. The electrical current flows in a geometric path contributing to the conductivity, which is assumed to be the same for these two conductance elements. The geometric path is described by the intrinsic formation factor for shaly sand, F^*.

$$C_o = (1/F^*) \times (C_w + C_{clay}) \tag{5}$$

Clay conductance is described by the product of effective clay concentration, Q_v and the equivalent conductance of counterions, B. This model (Eq. 6) has been used for many years and proven successful for improving water saturation estimation in shaly sand.

$$1/R_t = F^* S_w^{(n^*)} (1/R_w + (BQ_v)/S_w) \tag{6}$$

Waxman-Smits equation conveniently reduced back to Archie equation when clay conductance is zero as in clean sand. Formation factor, F^* is the intrinsic value of the formation and carries the correlation to porosity as in Archie law (Eq. 3) (Hill and Milburn 1956). In the Waxman-Smits' method to determine F^*, a core plug is saturated repeatedly with brine (sodium chloride) of different salinity until the core plug's conductivity, C_o increase linearly with the brine's conductivity, C_w. F^* and m^* are obtained from the slope of linear part of the plot C_o versus C_w whereby slope $= 1/F^* = \emptyset^{m^*}$ (Fig. 1).

In the standard core analysis procedure, the value of F^* is determined from the equation $F^* = Fa(1 + BQ_vR_w)$. The apparent formation factor, Fa is increased by a factor of $(1 + BQ_vR_w)$ to account for the effect of clay surface conductivity in the measured value of Fa. F^* is then plotted versus porosity, $\emptyset$ on a log-log scale. m^* is the slope of graph based on Archie's law that $F^* = 1/\emptyset^{m^*}$.

The abovementioned core analysis procedure tried to produce intrinsic value of formation factor, which should be equal to Archie-type formation factor using linear correlation. This is particularly valid for every brine conductivity if B and Q_v are also changed with the changes in brine conductivity. An average value of m^* is normally obtained from the graphical plot of F^* versus porosity, $\emptyset$. However, the accuracy of

Fig. 1 Typical plot for C_o versus C_w

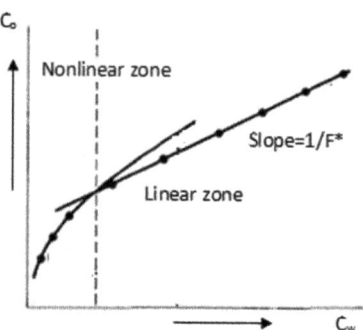

this method depends on (1) how representative the selected plugs to the reservoir are and (2) the accuracy of the parameters F and Q_v derived from the core analysis.

m* obtained from regression of graphical plot often resulted in higher value than the range 1.8–2.0 suggested by Archie. The higher value is believed to be caused by higher cementation in shaly sand but there could be other issues contributing to possible inaccuracy. The core plugs chosen for core analysis might vary in the properties especially in a heterogeneous reservoir. Average m* obtained from data regression of these core plugs might not be representative for the whole reservoir since different value of m* is expected in reservoir with different properties i.e. grain shape, type and size, shape and size of pores and pore throats, the size and numbers of dead-end pores (Salem and Chilingarian 1999). The influence of grain shape to the value of m* was also reported by others (Atkins and Smith 1961; Wyllie and Gregory 1953) and a value of 1.5 for perfect spheres was reported by Revil et. al. (1998) after Sen et al. (1981). Fixing m* in Waxman-Smits equation to a single value and using it to determine S_w in shaly sand could be misleading and this is often happened in practice whereby higher S_w is calculated in clean sand (as compared to using Archie equation with m value of 1.8 or 2.0). This would not happened if m* is allowed to change especially in clean sand.

In the core analysis procedures core plugs are subject to cleaning and drying. Possible alteration to the electrochemical properties of clay minerals might occur from removal of clay particles from reservoir rock during core cleaning process (Ali et al. 2011). The conductivity of clay minerals not only rendered by the amount of exchange cations but also the morphology and distribution of the clay (de Lima and Sharma 1990; Diederix and Yuan 1989). The Q_v measurement on pulverized samples might not be representative of the true Q_v as broken bonds had introduced additional exchangeable cations (Thomas 1976), which results in higher clay conductivity.

Accurate value of m* used in the S_w interpretation is important as this will impact the in-place estimation. Jorden (1971) pointed out that uncertainty in the value of m* could contribute 80–120% of errors in hydrocarbon volume estimates. m* contains the geometrical information on how the electrical current flows in the formation and this is also related to both ionic conduction in pore electrolyte and also conduction on clay mineral surface. The key to solve for more accurate m* is by understanding the mechanism of both conduction.

Ionic-surface Conduction in Shaly Sand and Induced Polarization.

The conductance in fully saturated shaly sand has been modelled by Waxman and Smits as consist of two parallel elements; one conductance element is contributed by the electrolyte in pore spaces and the other is due to the surface conductance of clay minerals. The clay conductance can be computed from the product of its cation exchange capacity (CEC) expressed as Q_v and the ionic conductance of the exchange cations, B. Electrical current in both conduction flows through the same geometrical path, F*.

$$C_o = 1/F^*(C_w + BQ_v) \tag{7}$$

The conduction in shaly sand can be explained from the clay structure, cation exchange process at the clay lattice and the ionic-surface conduction theory. We first described porous medium as proposed by Revil and Glover (1997) as consists of insulating grain and interconnected pore volume saturated in electrolyte. In shaly sand, clays can exist by means of dispersion in the pore bodies or coating and lining on the grain surfaces with different orientations depending on the clay types (de Lima and Sharma 1990; Thomas 1976).

Clay minerals are phyllosilicates with tetrahedral layer and octahedral layer as the basic building block. The tetrahedral layer consists of either Si or Al bonding with the basal oxygen in tetrahedral coordination while the octahedral layer consists of cations (Si, Al or Mg) bonding with the basal oxygen in octahedral coordination. The alumino-silicates minerals appear as layering either in 2:1 or 1:1 ratio. Smectite and illite are examples of 2:1 phyllosilicates (2 tetrahedral and 1 octahedral layer) and kaolinite is a 1:1 phyllosilicate. The structure of the clays that determine its physical and electrochemical properties (Kloprogge 1998).

Isomorphous substitution by the lower valence cations occurs at the tetrahedral sites; normally Si^{4+} by Al^{3+} and at the octahedral sites; Al^{3+} by Mg^{2+} or Mg^{2+} by Li^+. This introduces positive charge deficiency at the clay surface, which is balanced by the counterions such as Na^+, K^+ or Ca^{2+}. The cation exchange can also occur at the broken bonds around the crystal edge of the silica-alumina units that produce negative charge, which is also balanced by the hydrated interlayer cations (Revil et al. 1998) and by the ionized hydrogen of exposed surface hydroxyls (Ma and Eggleton 1999). The counterions are adsorbed on the clay minerals surface and can be exchanged for other cations.

The number of adsorbed counterions is the measures of clay minerals cation exchange capacity (CEC). The CEC of clay minerals depend on its structure, crystallinity and the clay geometry with respect to the fluid. For example, the cation exchange of montmorillonite occurs mostly on the basal plane surfaces giving high cation exchange capacity (CEC) of 80–150 meq/100 g.

In the theory of ionic-surface electrical conduction by Revil and Glover, the adsorbed counterions form a thin diffuse layer on the clay surface and contribute considerably to the effective electrical conductivity. The adsorbed water at mineral

Fig. 2 Complex conduction
model of shaly sand by
Vinegar and Waxman (1984)

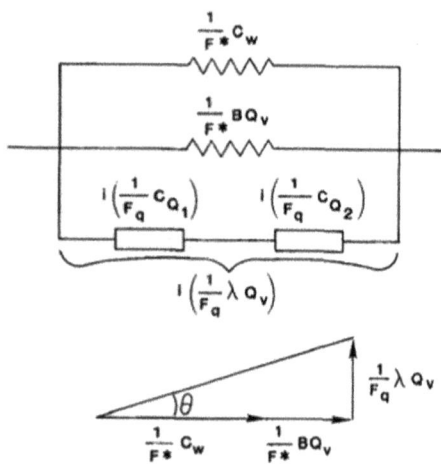

surface forms Stern layer and the combination with diffuse layer is called the electrical double layer. The ions in the diffuse layer is linked to the mineral surface through the Coulombic forces.

Stern layer and diffuse layer in the so-called electrical double layer contain counterions that balanced fixed charge on clay mineral surface. The ions in electrical double layer will polarize under alternating electrical field. An excess and deficiencies of ions at the clay sites will occur in steady state after a constant current being applied. A concentration gradient of ions will be developed due to clay counterions displacement on the clay surface and the electrolyte blockage by cations-selective membranes (Vinegar and Waxman 1984). The Maxwell-Wagner polarization caused by formation of field-induced free charge distribution at the interface of different phases of the medium is another polarization type occurs in porous media. The dispersion phenomena at low frequency is mainly controlled by polarization in Stern layer but can be overlapped by Maxwell-Wagner polarization in the frequency domain (Leroy and Revil 2009). As reported by Leroy and Revils, models incorporating the polarization phenomena into shaly sand conductivity were also developed by others (de Lima and Sharma 1992; Leroy et al. 2008; Cosenza et al. 2008).

The induced polarization (IP) is also known as the quadrature conduction at low frequency caused by the concentration gradient at the clay surface. The quadrature conduction is incorporated in Vinegar and Waxman complex conduction model of shaly sand (Fig. 2).

Vinegar and Waxman (1984) had proposed that imaginary part of the quadrature conduction is a function of shaliness or BQ_v and the quadrature equivalent conductivity having weak dependency to brine conductivity. As opposed to this, Revil and Skold (2011) reported that quadrature conductivity increases with salinity of the brine for value below 1 Sm^{-1}. Weller et al. (2013) used the dependency of the quadrature conductivity, C_Q as well as the clay surface conductivity, C'_{surf} to salinity to proposed a linear relationship between those two parameters:

$$1 = C_Q/(C'_{surf}) = 0.042 \tag{8}$$

The electrolyte conductivity, C_w and the clay mineral surface conductivity, C'_{surf} are the elements measured by the normal logging tool while the imaginary part of the surface conductivity, C_Q which is a result of charge storage can be measured by Induced Polarization (IP) tool. The relationship allows determination of clay surface conductivity from the complex conductivity measurement. The formation factor, F^* can be predicted from the following complex conductivity, C^* equations;

$$[[C]]^* = C' + iC_Q = 1/F^*C_w + C'_{surf} + iC_Q \tag{9}$$

$$F^* = C_w/(C' - C'_{surf}) = C_w/(C' - C_Q/1) \tag{10}$$

We elaborate the equation further to find m^*.

$$m^* = (\log(C' - C_Q/1) - \log C_w)/(\log \emptyset) \tag{11}$$

For this equation to be valid the polarization and has to change with changes in electrolyte conductivity. This allows the prediction of m^* from direct measurement of clay effect by means of induced polarization at any fluid salinity. The clay surface conductivity is predicted from the quadrature conductivity allowing it to be extracted from the total measured in-phase conductivity. This model requires complex conductivity measurement thus the application to the log interpretation depends on correlations between IP and well log parameters based on scientific laboratory research.

Conclusion

It is apparent that the induced polarization could provide a direct indication of electrochemical activities at clay mineral surface. Based on this it has been reported that IP concept can also be applied to estimate surface conductivity of clay mineral. Estimation or quantification of the clay mineral conductivity could lead to quantification of other important electrical properties of formation especially in shaly sand whereby most of the well logging data has been masked by the presence of clay. Determination of formation factor, F^* and m^* could be more accurate using the IP concept and model proposed in this paper. It seems that the way forward is to realize the model using proper correlation of IP and well log data from scientific laboratory research.

Acknowledgements The author acknowledges a sponsorship of the program from Ministry of Higher Education (MOHE), Malaysia, together with Universiti Teknologi MARA (UiTM), Yayasan UTP (YUTP) of Universiti Teknologi PETRONAS for funding the research project and Mr. Chiew Fook Choo from PETRONAS Carigali Sdn Bhd for technical advices and support.

References

Ali, Arfan, et al. 2011. Quantifying the Effects of Core Cleaning, Core Flooding and Fines Migration Using Sensitive Magnetic Techniques: Implications for Permeability Determination and Formation Damage. *Petrophysics* 52 (06): 444–451.

Archie, Gustave E. 1942. The Electrical Resistivity Log as an Aid in Determining Some Reservoir Characteristics. *Transactions of the AIME* 146 (01): 54–62.

Atkins Jr., E.R., and G.H. Smith. 1961. The Significance of Particle Shape in Formation Resistivity Factor-Porosity Relationships. *Journal of Petroleum Technology* 13 (03): 285–291.

Cosenza, Philippe, et al. 2008. A Physical Model of the Low-frequency Electrical Polarization of Clay Rocks. *Journal of Geophysical Research: Solid Earth* 113 (B8).

de Lima, Olivar A.L., and Mukul M. Sharma. 1990. A Grain Conductivity Approach to Shaly Sandstones. *Geophysics* 55 (10): 1347–1356.

de Lima, Olivar A.L., and Mukul M. Sharma. 1992. A Generalized Maxwell-Wagner Theory for Membrane Polarization in Shaly Sands. *Geophysics* 57 (3): 431–440.

Diederix, K.M., and H.H. Yuan. 1989. The Role of Membrane Potential Measurements in Shaly Sand Evaluation. *The Log Analyst* 30 (06).

Herrick, D.C., and W. David, Kennedy. 1993. Electrical Efficiency: A Pore Geometric Model for the Electrical Properties of Rocks. In SPWLA 34th Annual Logging Symposium. Society of Petrophysicists and Well-Log Analysts.

Herrick, David C., and W. David Kennedy. 1994. Electrical Efficiency—A Pore Geometric Theory for Interpreting the Electrical Properties of Reservoir Rocks. *Geophysics* 59 (6): 918–927.

Herrick, D.C., and W.D. Kennedy. 2009. A New Look at Electrical Conduction in Porous Media: A Physical Description of Rock Conductivity. SPWLA 50th Annual Logging Symposium. Society of Petrophysicists and Well-Log Analysts.

Hill, H. Jo., and J.D. Milburn. 1956. Effect of Clay and Water Salinity on Electrochemical Behavior of Reservoir Rocks. *Petroleum Transactions* 207: 65–72.

Jorden, J.R. 1971. Goals for Formation Evaluation. *Journal of Petroleum Technology* 23 (01): 55–62.

Kennedy, W. David, and David C. Herrick. 2012. Conductivity Models for Archie Rocks. *Geophysics* 77 (3): WA109–WA128.

Kloprogge, J.T. 1998. Synthesis of Smectites and Porous Pillared Clay Catalysts: A Review. *Journal of Porous Materials* 5 (1): 5–41.

Leroy, Philippe, and André Revil. 2009. A Mechanistic Model for the Spectral Induced Polarization of Clay Materials. *Journal of Geophysical Research: Solid Earth* 114 (B10).

Leroy, Ph, et al. 2008. Complex Conductivity of Water-Saturated Packs of Glass Beads. *Journal of Colloid and Interface Science* 321 (1): 103–117.

Ma, Chi, and Richard A. Eggleton. 1999. Cation Exchange Capacity of Kaolinite. *Clays and Clay Minerals* 47 (2): 174–180.

Owen, J.E. 1952. The Resistivity of a Fluid-Filled Porous Body. *Journal of Petroleum Technology* 4 (07): 169–174.

Patnode, H.W., and M.R.J. Wyllie. 1950. The Presence of Conductive Solids in Reservoir Rocks as a Factor in Electric Log Interpretation. *Journal of Petroleum Technology* 2 (02): 47–52.

Revil, A., and P.W.J. Glover. 1997. Theory of Ionic-Surface Electrical Conduction in Porous Media. *Physical Review B* 55 (3): 1757.

Revil, André, and Magnus Skold. 2011. Salinity Dependence of Spectral Induced Polarization in Sands and Sandstones. *Geophysical Journal International* 187 (2): 813–824.

Revil, A., et al. 1998. Electrical Conductivity in Shaly Sands with Geophysical Applications. *Journal of Geophysical Research: Solid Earth* 103 (B10): 23925–23936.

Salem, Hilmi S., and George V. Chilingarian. 1999. The Cementation Factor of Archie's Equation for Shaly Sandstone Reservoirs. *Journal of Petroleum Science and Engineering* 23 (2): 83–93.

Sen, P.N., C. Scala, and M.H. Cohen. 1981. A Self-similar Model for Sedimentary Rocks with Application to the Dielectric Constant of Fused Glass Beads. *Geophysics* 46 (5): 781–795.

Thomas, E.C. 1976. The determination of Qv from Membrane Potential Measurements on Shaly Sands. *Journal of Petroleum Technology* 28 (09): 1–87.

Vinegar, H.J., and M.H. Waxman. 1984. Induced Polarization of Shaly Sands. *Geophysics* 49 (8): 1267–1287.

Waxman, Monroe H., and L.J.M. Smits. 1968. Electrical Conductivities in Oil-Bearing Shaly Sands. *Society of Petroleum Engineers Journal* 8 (02): 107–122.

Weller, Andreas, Lee Slater, and Sven Nordsiek. 2013. On the Relationship Between Induced Polarization and Surface Conductivity: Implications for Petrophysical Interpretation of Electrical Measurements. *Geophysics* 78 (5): D315–D325.

Wildenschild, Dorthe, Jeffery J. Roberts, and Eric D. Carlberg. 2000. On the Relationship Between Microstructure and Electrical and Hydraulic Properties of Sand-Clay Mixtures. *Geophysical Research Letters* 27 (19): 3085–3088.

Worthington, Paul F. 1985. The Evolution of Shaly-Sand Concepts in Reservoir Evaluation. *The Log Analyst* 26(01).

Wyllie, M.R.J., and A.R. Gregory. 1953. Formation Factors of Unconsolidated Porous Media: Influence of Particle Shape and Effect of Cementation. *Journal of Petroleum Technology* 5 (04): 103–110.

Experimental Studies of the Effect of Ionic Strength on Epoxy-Based Polymer for Water Shut-off Operation

Hasian P. Septoratno Siregar, Taufan Marhaendrajana, Priozky P. Purba, Wingky Suganda, William Angtony, Michael Y. Suryana and Kharisma Idea

Abstract This paper provides the experimental studies on parameters affecting the performance of epoxy based polymer used to be injected to unproductive layers in a water shut-off application. In this paper, we observe a parameter that can affect the performance of epoxy-based polymer which is ionic strength. By focusing the research on epoxy-based polymer, this study tackles the environmental problem and operation cost in water shut-off operation by proposing more environmental friendly and much cheaper polymer. The epoxy-based polymer can replace the use of the conventional Cr(III)-Carboxylate/Acrylamide-Polymer (CC/AP) which is more expensive and creates environmental problems. The epoxy-based polymer is tested at various ionic strength to determine the effect of ionic strength on density, rheological properties, gelation time and hard gel compressive strength after the polymer has gelled and hardened. The density of the polymer is determined analytically. The mass of the polymer is determined by weighing the gel, and the volume of the polymer was measured using the Archimedes method that measures the volume of irregularly shaped object to measure the gel volume. The density of the polymer is then calculated as the mass of the gel divided by the volume of the gel. To determine the rheological properties, the epoxy based polymer tested on a rotational viscometer at various time. Then, we observe the gelation time of epoxy-based polymers with semi-quantitative method by comparing the gel strength development with gel strength code US Patent No. 4688639. After it has gelled and hardened, the polymer is tested on hydraulic press equipment to determine the compressive strength of the polymer.

Keywords Epoxy-based polymer · Water shut-off · Ionic strength

H. P. S. Siregar (✉) · T. Marhaendrajana · P. P. Purba · W. Suganda · W. Angtony ·
M. Y. Suryana · K. Idea
Institut Teknologi Bandung, Bandung, Indonesia
e-mail: septo@tm.itb.ac.id

© Springer Nature Singapore Pte Ltd. 2018
B. M. Negash et al. (eds.), *Selected Topics on Improved Oil Recovery*,
https://doi.org/10.1007/978-981-10-8450-8_2

Introduction

As we know, nowadays, the upstream sector of the oil and gas industry has been on depletion phase since oil and gas have been produced for many decades. On the contrary, the demand for oil and gas increase incessantly, so it becomes a challenge for us to meet the demand of the world's energy needs, especially oil and gas, amid production of oil and gas which has reached the depletion stage. One of the solutions to address this challenge is the cutting-edge technology in production of oil and gas, called Enhanced Oil Recovery.

One of the problems that is highlighted in this work is the problem of excessive unwanted water production in association with crude oil production. This is one of the major difficulties for the petroleum industry, as more reservoirs become mature. This water production often decreases the economic life of a well because of problems such as increasing corrosion rates, increasing tendency for emulsion, scale formation, environmental concerns, coning due to bottom water drive, and increasing production cost of lifting, handling, separation, and disposal of the produced water. The solution that we propose to overcome these issues is the use of chemical EOR (Enhanced Oil Recovery) by injecting gel polymer to the fractures or high permeability zones that produce a lot of water. This method will shut-off the water production by blocking the porosity and reducing the permeability of those zones in the reservoir. So, well treatments using polymer is attractive, especially for wells with complicated completion design (deviated, horizontal, multilateral).

The polymer gel systems also have several important advantages compared with other methods: (1) The polymer system, that is injected as a solution, can penetrate deep in the reservoir and reduce the permeability in the near wellbore area; (2) The injected solution can move up and down outside the wellbore, sealing cracks and existing micro-annuli especially in poorly cemented high permeable sands; (3) Inexpensive due to reduced crew and rig time. In addition, epoxy has already been commonly used in completion operation for various purposes, such as treating observation wells with glass reinforced epoxy pipes to obtain higher accuracy and a deeper investigation of the formation saturation (Mukmin et al. 2010) and controlling sand and water production (Brooks et al. 1974).

According to previous study by Hakiki et al. (2015a, b) it was shown that epoxy-based polymer has big potential to be used as chemical injection that can be applied in water shut-off application. This research focuses on the effect of ionic strength to the performance of epoxy-based polymer for water shut-off application. We will vary the independent parameter ionic strength of this polymer solution. Then, we will observe the dependent parameter such as gelation time, density, rheology parameter, and compressive strength.

Methodology

Varying the Ionic Strength Content of Epoxy-Based Polymers

Before conducting the research, four cups of epoxy-based polymer are prepared. These polymer samples were tested with several different values of ionic strength. In order to vary the ionic strength values, the samples are mixed samples with different type of salt (especially NaCl, KCl, $CaCl_2$, $MgCl_2$). The salt concentration in the solutions is maintained at 1 Molar, by adjusting the total mass of salt added into the solution.

Observing the Density of the Epoxy-Based Polymers

The densities of the samples are measured before and after it hardens into a gel. The densities of the sample are determined using analytical calculation by using the volumes and masses of the hardened epoxy-based polymers. The volumes of the hardened epoxy-based polymers are measured using the volumetric method (Fig. 1).

Observing the Rheological Properties of Epoxy-Based Polymer

The rheological properties of the samples are measured using rotational viscometer to determine the development of their properties with time (Fig. 2).

Observing the Gelation Time of Epoxy-Based Polymers

The measurement of gelation time is carried out using the bottle testing method, which is a semi quantitative method to determine the gelation time by comparing the gel strength development with gel strength code US Patent No. 4688639. The gelation time is obtained when the gel is unable to flow in the bottle anymore and there is no gel surface deformation by gravity upon inversion. The observation is conducted by reversing the bottle position.

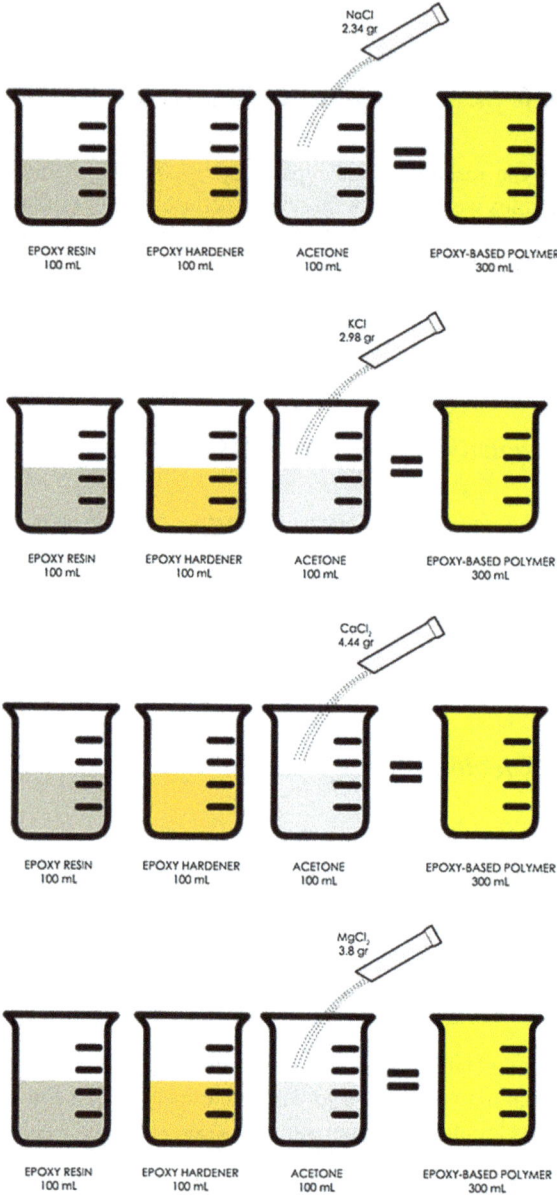

Fig. 1 The making of epoxy-based polymer with varying ionic strength

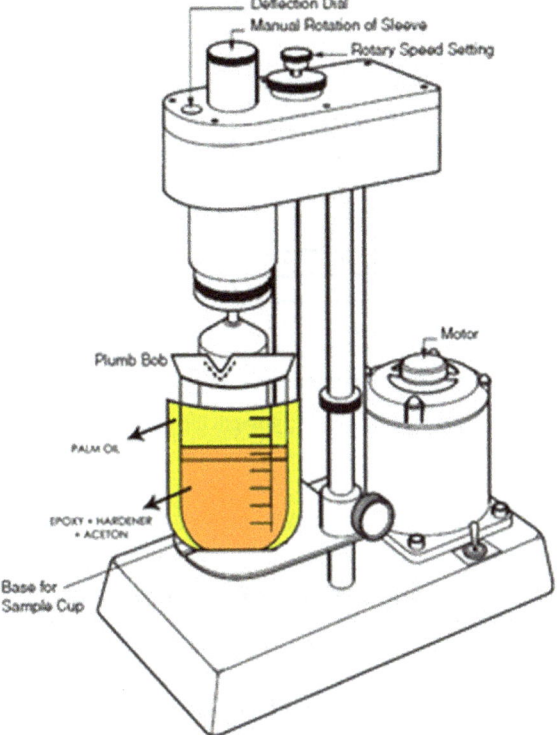

Deflection Dial
Manual Rotation of Sleeve
Rotary Speed Setting

Motor

Plumb Bob

PALM OIL

EPOXY + HARDENER
+ ACETON

Base for
Sample Cup

Fig. 2 The schematic of measuring the rheological properties of the epoxy-based polymers

Observing the Compressive Strength of Epoxy-Based Polymer

After the polymer has hardened, its compressive strength is determined using hydraulic press. This apparatus provide pressure hydraulically on block bearing, which will press the sample at certain pressure. In order to determine compressive strength, the pressure when the sample begin to crack is noted (Figs. 3, and 4).

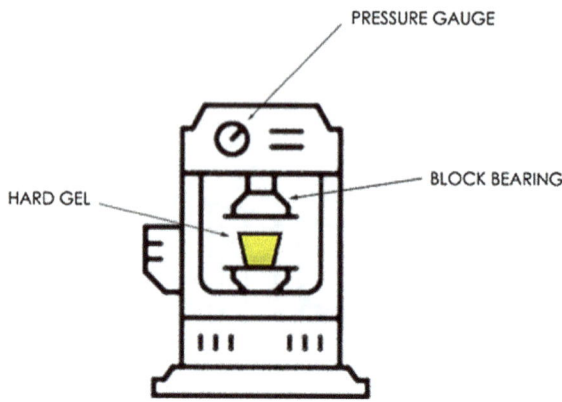

Fig. 3 The schematic of measuring compressive strength the hardened epoxy-based polymer

Result and Discussion

The Variations of the Ionic Strength Content of Epoxy-Based Polymer

The Effect of Ionic Strength Content on Density

By applying the above methodology, the following table can be generated (Table 1).

It can be seen that the density of epoxy-based polymers has increased during the gelation period. That phenomenon can happen because epoxy-based polymers is densified during the gelation time. Also, it can be inferred that the different ionic strength content in epoxy-based polymers does not significantly affect the density of the gel.

Table 1 Density of initial gel and hard gel for ionic strength experiment

Sample	Density (g/ml)	
	Initial gel	Hard gel
NaCl	0.798	1.124
KCl	0.798	1.125
$MgCl_2$	0.798	1.123
$CaCl_2$	0.798	1.124

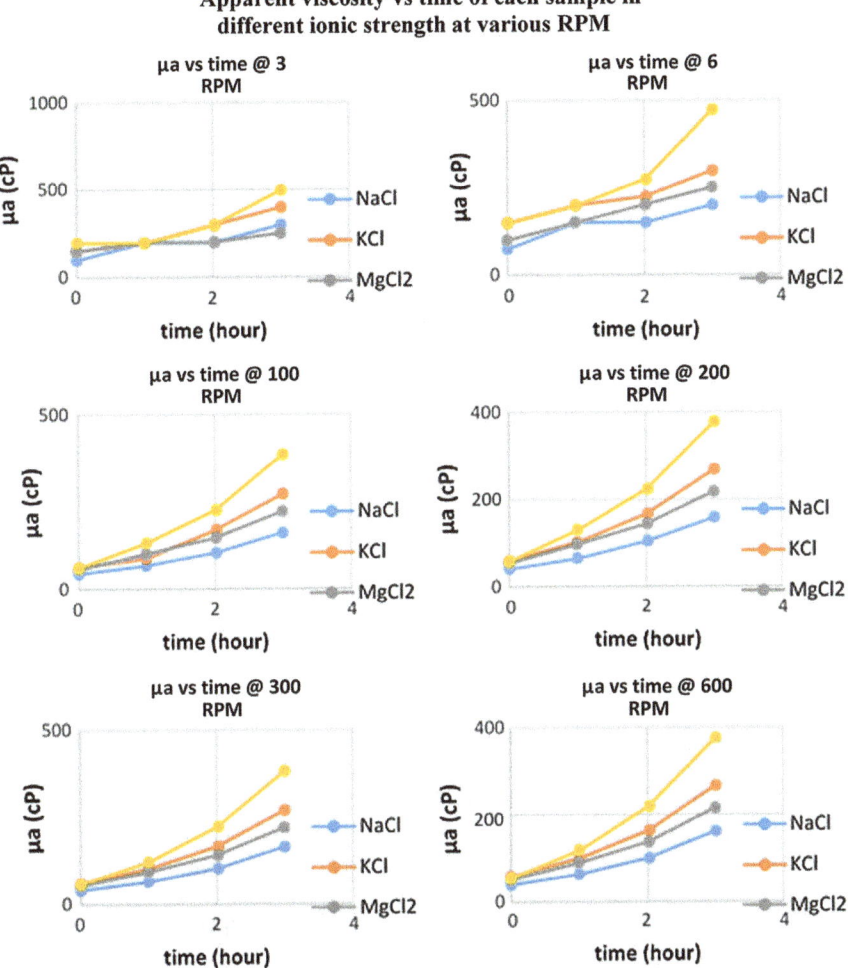

Fig. 4 Apparent viscosity versus time of each sample at various RPM for ionic strength experiment

The Effect of Ionic Strength Content on Rheological Properties

By applying the above methodology, the following is the table showing the result of dial reading by using Rotational viscometer for this section (Table 2).

By using Rotational viscometer calculations, the dial reading is converted to various rheological properties, then the following tables can be generated, (Tables 3, 4 and 5).

From the tables above, the following graphs can be constructed.

From laboratory data processing above, the increasing ionic strength content in epoxy-based polymers increase the overall rheological properties of epoxy-based

Table 2 The result of dial reading by using rotational viscometer for ionic strength experiment

Sample	Time	Dial reading (8)					
	(h)	3 RPM	6 RPM	100 RPM	200 RPM	300 RPM	600 RPM
ER + EH + Act + NaCl	0	1	1.5	15	28	41	79
	1	2	3	23	44	66	129
	2	2	3	35	70	103	202
	3	3	4	54	106	165	326
ER + EH + Act + KCl	0	1.5	3	21	41	61	118
	1	2	4	30	69	104	203
	2	3	4.5	57	112	167	329
	3	4	6	91	180	270	535
ER + EH + Act + MgCl$_2$	0	1.5	2	19	36	53	100
	1	2	3	34	65	93	180
	2	2	4	49	96	140	274
	3	2.5	5	74	145	219	430
ER + EH + Act + CaCl$_2$	0	2	3	21	40	59	111
	1	2	4	45	88	124	241
	2	3	5.5	76	150	224	440
	3	5	9.5	129	253	382	754

polymers. It can be concluded that all of epoxy-based polymers in different ionic strength follow the fluid behaviour of Bingham Fluid, since it can be seen from Fig. 5 that all samples will show linear relationship between shear stress and shear rate and also since from Fig. 6, all samples will give nonzero value of gel strength.

From Fig. 6, it also can be concluded that all of epoxy-based polymers in different ionic strength are low-flat gels since its show low and constant gel strength development during the gelation time. Low-flat gel is desirable when injecting the polymers down the well since in this type of gel, the gel strength of sample that is left in static condition for a long time will not be much more viscous than the initial condition. Thus, epoxy-based polymers have relatively long time of pumpable state although it is left in static condition for a relatively long time (Fig. 4).

The Effect of Ionic Strength Content on Gelation Time

By comparing with the reference from US Patent No. 4688639 on gel strength code, we observe the gel strength development of each sample epoxy-based polymer every hour since it is mixed until it meets the gelation time criteria. The following is a graph showing the gel strength development every hour of each epoxy-based polymer with different ionic strength. Figure 7 shows the effect of ionic strength content on gelation time.

Table 3 The tabulation of rheological properties calculation for ionic strength experiment

Sample	Time (h)	PV (cp)	Yield point (lb/100 ft^2)	Gel strength (lb/100 ft^2)	Apparent viscosity (cp)			
					3 RPM	6 RPM	100 RPM	200 RPM
ER + EH + Act + NaCl	0	38	3	1	100	75	45	42
	1	63	3	2	200	150	69	66
	2	99	4	2	200	150	105	105
	3	161	4	3	300	200	162	159
ER + EH + Act + KCl	0	57	4	1.5	150	150	63	61.5
	1	99	5	2	200	200	90	103.5
	2	162	5	3	300	225	171	168
	3	265	5	4	400	300	273	270
ER + EH + Act + MgCl$_2$	0	47	6	1.5	150	100	57	54
	1	87	6	2	200	150	102	97.5
	2	134	6	2	200	200	147	144
	3	211	8	2.5	250	250	222	217.5
ER + EH + Act + CaCl$_2$	0	52	7	2	200	150	63	60
	1	117	7	2	200	200	135	132
	2	216	8	3	300	275	228	225
	3	372	10	5	500	475	387	379.5

The following is a table showing the gelation time of each epoxy-based polymer with different ionic strength (Table 6).

It can be seen that the gelation time of epoxy-based polymer added by NaCl is the slowest and the gelation time of epoxy-based polymer added with CaCl$_2$ is the fastest. According to the reference (Lewis and Randal 1921), the ionic strength of NaCl < KCl < MgCl$_2$ < CaCl$_2$. It can be inferred that the stronger ionic strength of salts, the gelation time will be shorter.

The Effect of Ionic Strength Content to Compressive Strength

By using Uniaxial Hydraulic Press calculations, the pressure gauge reading is converted to compressive strength then the following tables can be generated (Table 7).

The data above show that the compressive strength of epoxy-based polymers in the form of hard gel increase with increasing the ionic strength content in epoxy-based polymers.

Table 4 The tabulation of shear stress and shear rate calculation for ionic strength experiment

Sample	Time (h)	Shear stress (dyne/cm^2)					
		3 RPM	6 RPM	100 RPM	200 RPM	300 RPM	600 RPM
ER + EH + Act + NaCl	0	5.08	7.62	76.16	142.16	208.16	401.08
	1	10.15	15.23	116.77	223.39	335.08	654.93
	2	10.15	15.23	177.70	355.39	522.93	1025.55
	3	15.23	20.31	274.16	538.16	837.71	1655.10
ER + EH + Act + KCl	0	7.62	15.23	106.62	208.16	309.70	599.09
	1	10.15	20.31	152.31	350.31	528.01	1030.63
	2	15.23	22.85	289.39	568.62	847.86	1670.33
	3	20.31	30.46	462.01	913.86	1370.79	2716.20
ER + EH + Act + MgCl$_2$	0	7.62	10.15	96.463	182.77	269.08	507.70
	1	10.15	15.23	172.62	330.01	472.16	913.86
	2	10.15	20.31	248.77	487.39	710.78	1391.10
	3	12.69	25.39	375.70	736.17	1111.86	2183.11
ER + EH + Act + CaCl$_2$	0	10.15	15.23	106.62	203.08	299.54	563.55
	1	10.15	20.31	228.47	446.78	629.55	1223.56
	2	15.23	27.92	385.85	761.55	1137.25	2233.88
	3	25.39	48.23	654.93	1284.48	1939.41	3828.06
Shear rate (s^{-1})		5.11	10.22	170.40	340.80	511.20	1022.40

Table 5 The tabulation of gel strength for ionic strength experiment

Sample	Time (h)	GS 10 s	GS 10 min
ER + EH + Act + NaCl	0	1	2
	1	2	2
	2	2	3
	3	3	4
ER + EH + Act + KCl	0	2	2
	1	3	3
	2	3	4
	3	4	4
ER + EH + Act + MgCl$_2$	0	2	2
	1	3	3
	2	3	4
	3	4	5
ER + EH + Act + CaCl$_2$	0	3	4
	1	4	5
	2	5	5
	3	5	6

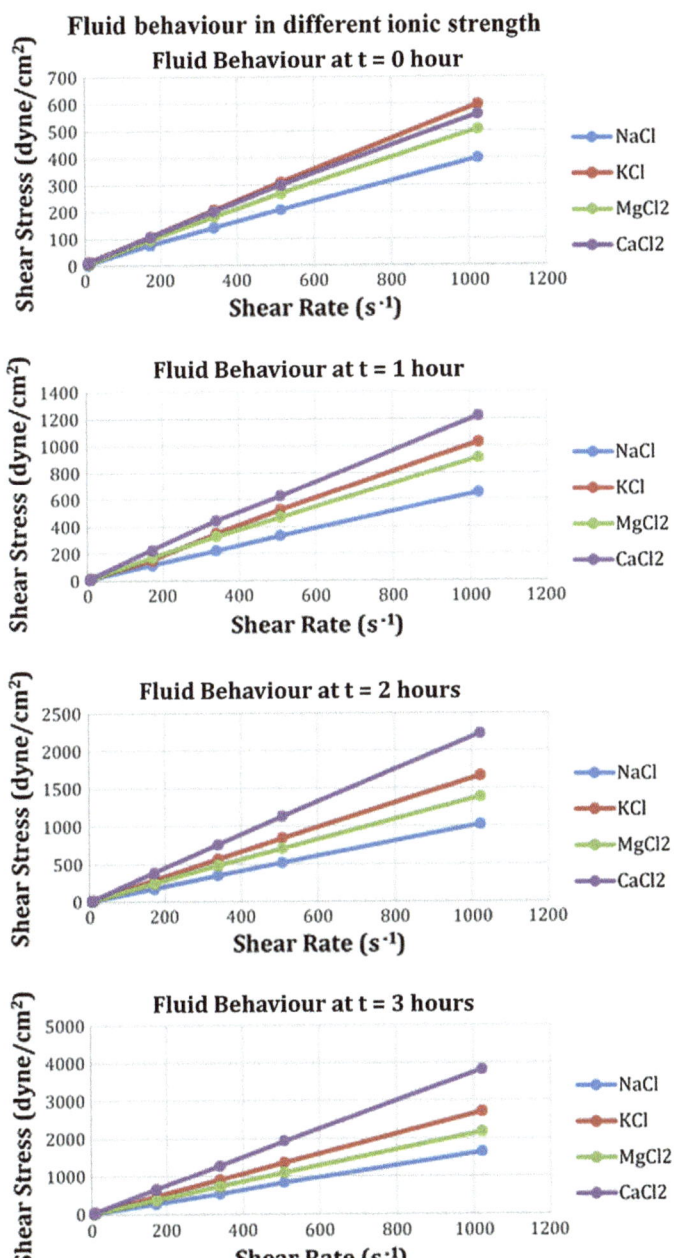

Fig. 5 Fluid behaviour of each sample for ionic strength experiment

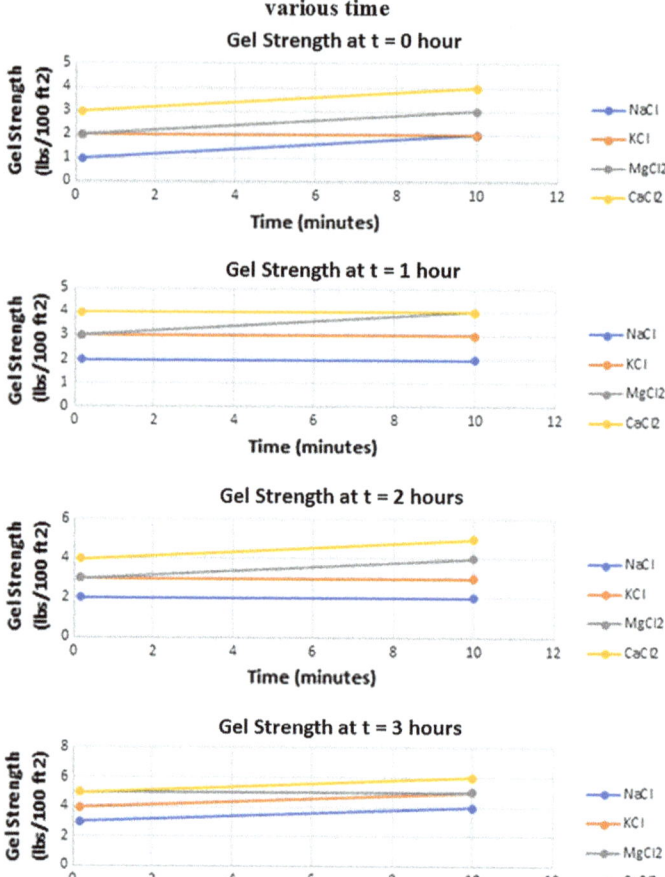

Fig. 6 Gel Strength of each sample for ionic strength experiment

Table 6 The gelation time of each epoxy-based polymer for ionic strength experiment

Sample	Gelation time (h)
ER + EH + Act + NaCl	29
ER + EH + Act + KCl	25
ER + EH + Act + MgCl$_2$	20
ER + EH + Act + CaCl$_2$	18

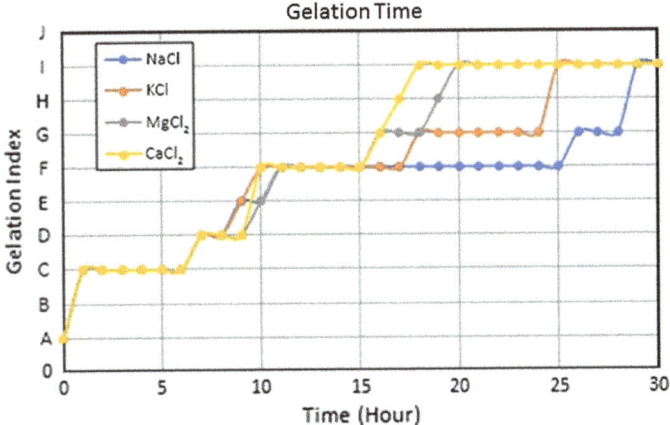

Fig. 7 The gelation time of each sample epoxy-based polymer for ionic strength experiment

Table 7 Compressive strength calculations for ionic strength experiment

Sample	Height (cm)	Diameter (cm)	t/d	Area (cm2)	Correction factor, k	ΔP (psi)	Compressive strength
NaCl	6.76	7.01	0.97	38.50	0.86	165	415.97
KCl	7.00	7.25	0.97	41.28	0.86	190	446.74
MgCl$_2$	6.75	6.83	0.99	36.62	0.87	210	561.88
CaCl$_2$	6.74	6.98	0.97	38.26	0.86	255	646.86

Conclusion

From this experimental study, we can conclude that the increasing ionic strength content in the epoxy-based polymers does not affect significantly the density of the epoxy-based polymers. Meanwhile, the increasing ionic strength content in the epoxy-based polymers increases the rheological properties and the compressive strength but will decrease the gelation time. It can be seen that the density of epoxy-based polymers increases during the gelation period. That phenomenon can happen because epoxy-based polymers is densified during the gelation time. For future work, we will observe the other properties of epoxy-based polymers such as permeability reduction, porosity reduction, wettability, and interfacial tension.

References

Brooks, F.A., et al. 1974. Externally Catalyzed Epoxy For Sand Control. *SPE Journal of Petroleum Technology*.

Hakiki, F., et al. 2015a. Is Epoxy-Based Polymer Suitable for Water Shut-Off Application? SPE-176457-MS. Presented at the The Asia Pacific Oil & Gas Conference and Exhibition (APOGCE).

Hakiki, F., et al. 2015b. Preliminary Study of Epoxy-Based Polymer for Water Shut-Off Application. Paper IPA15-SE-025. Presented at the Thirty-Ninth Annual Convention & Exhibition of Indonesian Petroleum Association, May 2015.

Lewis, G., and M. Randall. 1921. The Activity Coefficient of Strong Electrolytes. Journal of the American Chemical Society 43(5):1112–1154. http://dx.doi.org/10.1021/ja01438a014

Mukmin, M., et al. 2010. Polymer Trial Using Horizontal Wells: Conceptual Well Completion Design and Surveillance Planning Aspects. SPE EOR Conference at Oil and Gas West Asia, 16–18 April, Muscat, Oman.

Vaishnav, Manoj, et al. 2010. Study *The Effect of Different Parameters on The Gelation of An Organically Cross-Linked Polymer Gel System*. India: Department of Petroleum Engineering Indian School of Mines.

Reservoir Characterization of Lahat Outcrop for the Application of Chemical Flooding in Air Benakat Sandstone Reservoir, Center of Sumatera

Kharisma Idea, Wahyu Vian Pratama, Taufan Marhaendrajana, Sudjati Rachmat and IGB Eddy Sucipta

Abstract The Enhanced Oil Recovery (EOR) method is a widely used for increasing oil recovery, one of which is by injecting ionic surfactants in sandstone reservoirs. The Sandstone Reservoir in Benakat Air Formation is composed of grains with dominant quartz, and cement with dominant calcite and silica. Brown clay (montmorillonite) fill in the rock matrix. The mineral montmorillonite is highly reactive to water, so that the sodium in montmorillonite will be hydrated with surfactant solution and result in swelling which can result in shrinking of the pore size of the rock and decrease the permeability so that the surfactant injection results are not optimum. Clay minerals and calcite have a positive surface charge at the fluid pH conditions in the reservoir so that it will affect the degree of adsorption and or precipitation of the anionic surfactant. Such adsorption and precipitation may affect the incremental oil recovery by injection of surfactant solution in the Sandstone reservoir, Air Benakat formation, either positively or negatively. Reservoir characterization has been done by examining the mineral content of sandstones reservoir on thin section. Two types of outcrops were analyzed in this study which consist of two thin section (L1 and L2). The mineral content of L1 outcrop consists of quartz, Potassium feldspar, plagioclase, gluconite and fossils, where matrix is composed of quartz, potassium feldspar and clay with dominant calcite cement. The L2 outcrop consists of grains filled with quartz, glauconite, plagioclase, biotite and fossil, where matrix is dominated by quartz and then clay, with calcite cement. The objectives of this study are to characterize the minerals contents in the sandstone core of Lahat outcrop for consideration of surfactant injection in Air Benakat Sandstone reservoir.

Keywords Surfactant · Anionic · Clay · Montmorillonite · Calcite · Adsorption

K. Idea (✉) · T. Marhaendrajana · S. Rachmat
Petroleum Engineering Study Program, Institut Teknologi Bandung, Bandung, Indonesia
e-mail: kharismaidea@gmail.com

W. V. Pratama · I. Eddy Sucipta
Geological Engineering Study Program, Institut Teknologi Bandung, Bandung, Indonesia

© Springer Nature Singapore Pte Ltd. 2018
B. M. Negash et al. (eds.), *Selected Topics on Improved Oil Recovery*,
https://doi.org/10.1007/978-981-10-8450-8_3

Introduction

Meruap Field located in the South Sumatra Basin is one of the back arc basin of Sumatra regional tectonics occupying an area of 6433 km^2 (Fig. 1). Figure 2 shows the stratigraphy column of South Sumatra Basin. The formations in the Meruap blocks sequentially from old to young are Pre-Tertiary (BSM), Lahat Formation/Lemat Formation (LAF), Talang Akar Formation (TAF), Baturaja Formation (BRF), Gumai Formation (GUF), Air Benakat Formation (ABF), MuaraEnim Formation (MEF) and Tuff Kasai Formation (KAF). This study aimed to evaluate the application of chemical flooding in Air Benakat Formation which composed of sandstone reservoir. Air Benakat Formation was deposited during the regression phase and the end of deposition of Gumai formation in the middle miocene (Bishop 2001). The deposition in this regression phase occurs in the neritic environment until shallow marine, and then change into a delta plain and coastal swamp at the end of the first regression cycle.

This formation consists of grayish white clay with fine sandstone intercalation, bluish black grey sandstone, localized glauconite contains lignite and at the top containing tuff while abundance foraminifera in the middle of. The thickness of this formation is estimated between 1000–1500 m (Darman and Sidi 2000). This study investigates the rock minerals, such as montmorillonite and calcite that are present in reservoir rocks, which may affect the performance of ionic surfactant injection. The interaction with the surfactant and inorganic ion such as Na$^+$, Ca^{2+}, and Mg^{2+} that introduced in rock minerals can lead to precipitation and reservoir plugging (Somasundaran et al. 1984). There are five minerals (montmorillonite, calcite, dolomite, kaolinite and silica) were responsible for surfactant adsorption (Grigg et al. 2005). The reservoir characteristics of each reservoir affect the economy and success of the EOR method (Webb et al. 2012).

The mineral contents and grain composition of rocks were determined using thin section and petrographic analysis.

Methodology

Minerals analysis was performed using petrographic analysis under microscope observation on thin section using vary magnification of 5x, 10x, 20x, 50x and 100x. Two type of outcrop samples were utilized; L1 and L2 cores. Thin section preparation was done by adding blue day resin to observe porosity and alizarin red to observe the existance of calcite and dolomite minerals.

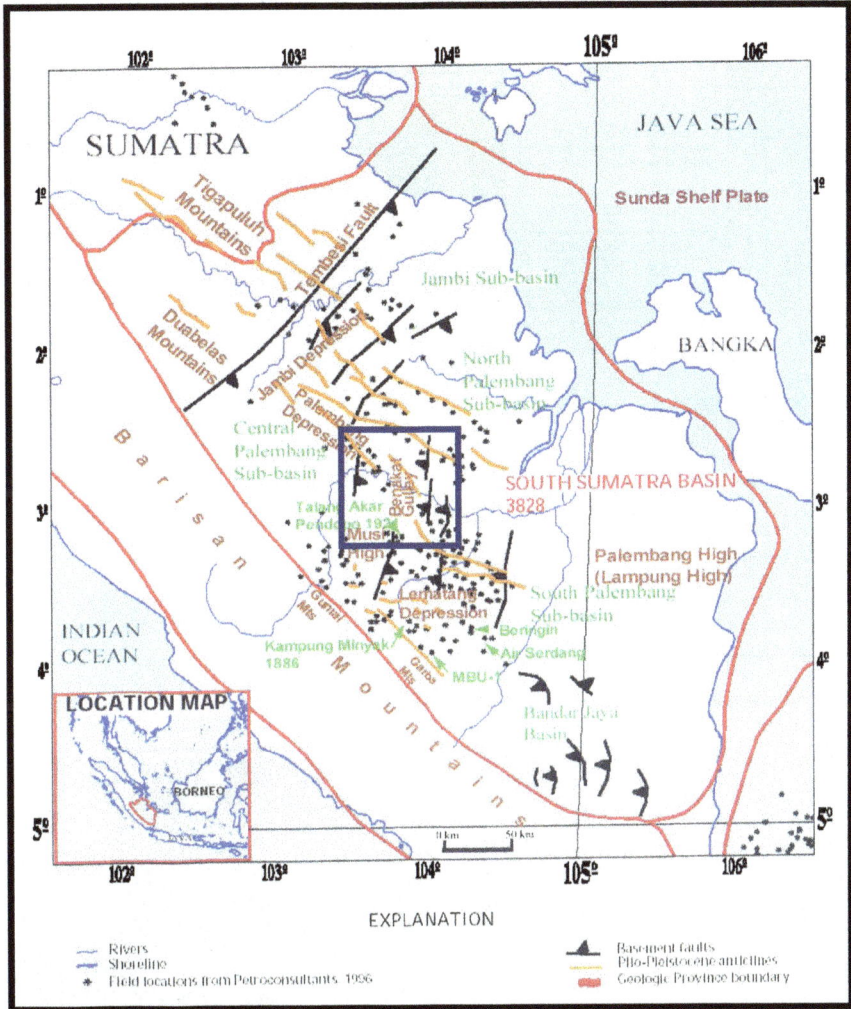

Fig. 1 Location of Air Benakat formation (Bishop 2001)

Analysis and Result

There were two types of sample analyzed in this study which consist of two thin sections. Both are calcareous sandstone and petrographically classified as Feldsphatic Wacke (Gilbert's classification in Williams et al. 1982). Based on petrographic analysis, the mineral content of sample L1 is consist of dominant quartz (12%), plagioclase (2%), Potassium feldspar (7%), muscovite (1%), glauconite (5%), fossil (5%) and lithic (3%), where matrix (30%) composed by quartz and Potassium feldspar with dominant calcite cement (25%). Porosity (10%) is intergranular and intragranular,

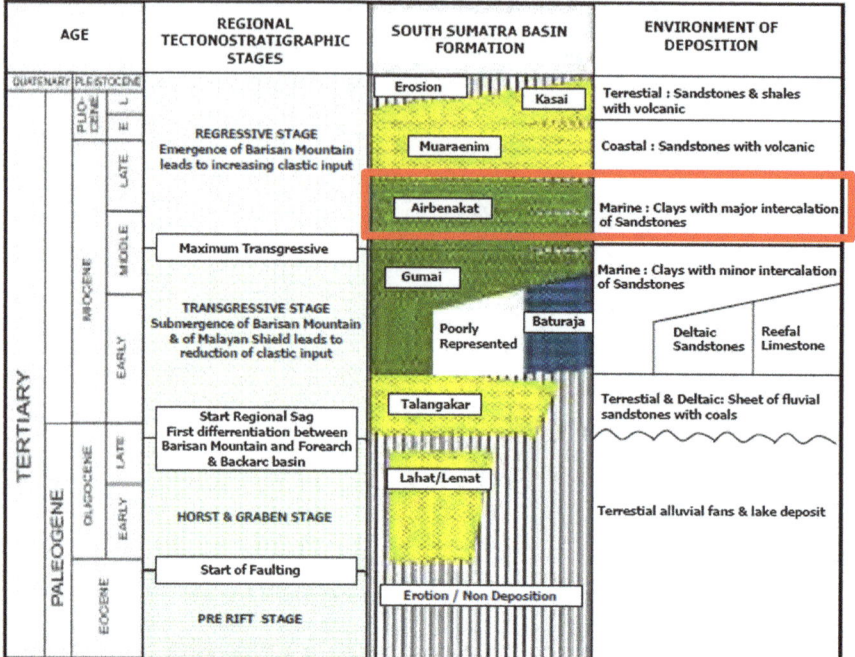

Fig. 2 Stratigraphy column of South Sumatra basin (Barber et al. 2005)

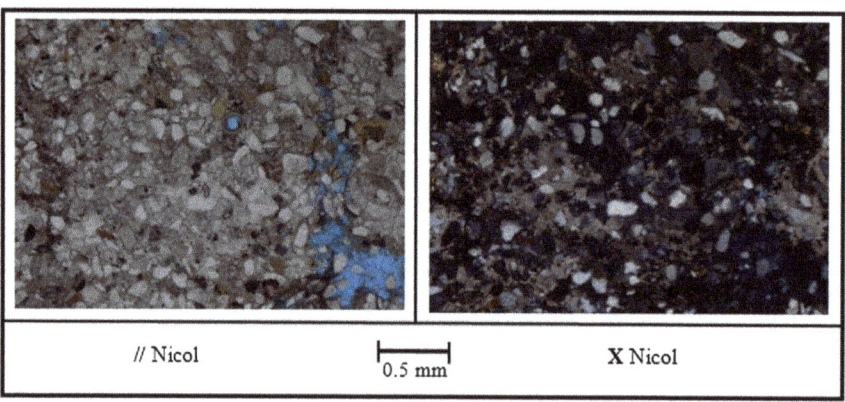

Fig. 3 Photomicrograph showing intergranular and intragranular porosity (sample L1-blueday)

localized due to dissolution (Fig. 3). Calcite in sample L1 may affect the injection of ionic surfactants which will allow the adsorption of surfactants in calcite minerals. The existance of calcite mineral can be seen in Fig. 4 where the red color indicates the calcite cement and the existance of clay mineral can be seen in Fig. 3 (X nikol) where the brown color around quartz grains indicates the montmorillonite.

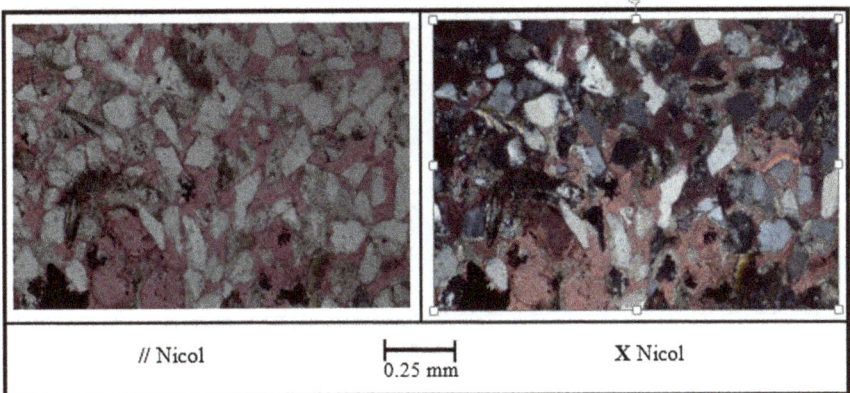

Fig. 4 Photomicrograph showing calcite cement with red colour (sample L1-alizarin red)

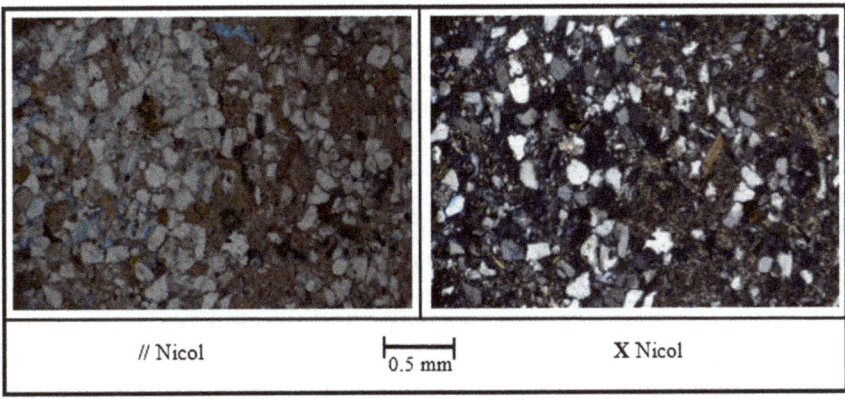

Fig. 5 Photomicrograph showing more organic matter and clay matrix (sample L2-blue day)

Sample L2 shows similar lithological characteristic with sample L1, but has less calcareoas content and more organic matter like foraminifera. Petrographically, sample L2 has clastic texture with moderately sorted. The grain is composed of quartz (10%), plagioclase (1%), Potassium feldspar (5%), biotite (1%), fossil (8%) and lithic (3%). The matrix consists of clay and dominant quartz (30%) with silica and calcite cement (25%) (Fig. 4). Porosity (10%) is formed by dissolution and dolomitization process as intergranular and intragranular porosity. Dolomitization process causes increasing porosity as carbonate mineral shrinkage (Fig. 6). The brown color around quartz grains indicates the montmorillonite where can be seen in Fig. 5 (X nikol) and Fig. 6 (X nikol).

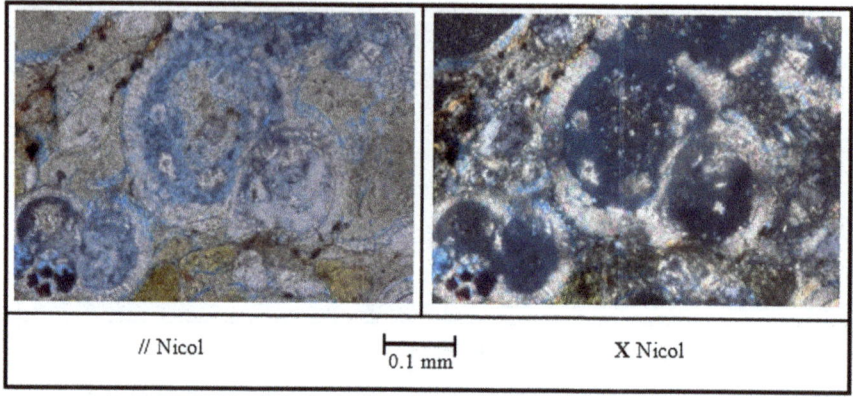

// Nicol   0.1 mm   X Nicol

Fig. 6 Photomicrograph showing dolomitization process which causes increasing porosity (sample L2-blue day)

Analysis of thin sections L1 and L2 compared with analysis of the core TPN #46 of the previous study from Marhaendrajana and Idea 2016. Figures 7 and 8 show the analysis of Air Benakat core (TPN #46) after surfactant injection. Air Benakat core was injected by surfactant and polymer. Surfactant concentration was 2% and 0.03% of polymer was added in the solution. There was no incremental recovery observed on Air Benakat core TPN #46. Core TPN #46 classified as greywacke sandstones. Petrographic analysis shows that core TPN #46 is clastic sedimentary rock with moderately grain sorted and point-long contact between grains. The grain framework is dominated by 65% quartz (D1), rounded—sub angular grain shape with matrix (15%) consists of clay and quartz (15%). Cement 15% (K6) consists of silica (10%) and carbonate (5%) with 5% porosity as intergranular. Carbonate cements fill the space between grains that leads to reduced porosity and cover the pores between grains that cannot remove the oil on the core (Fig. 7). Rock grains are arranged lengthwise so that the pores between the rocks were formed. Clay and carbonate cement that fills the pores between grains caused interlocking rock porosity. Secondary porosity was observed in the form of a cavity in the shell. This pores structure, grains arrangement, and cement filling cause trapped oil in the matrix and in the cavity of fossil shells, as they are covered by mineral clay and carbonate cement. This caused the injection of surfactant did not successfully displaced the oil that was trapped in the core TPN #46 (Fig. 8).

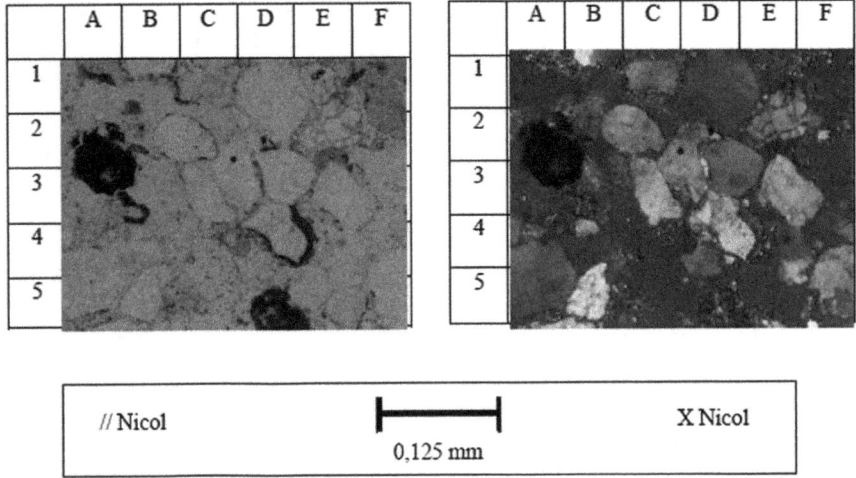

Fig. 7 Photomicrograph showing carbonate cement filling the space between grains, core sample of TPN #46

Conclusions

1. Reservoir characterization particularly on mineral contents aids to potentially detrimental effect during surfactant injection.
2. Swelling clay, precipitation calcite and compaction of grain reduced porosity and permeability.
3. Grain arrangement with interlocking porosity, secondary porosity and alignment of carbonates and clay cement makes trap residual oil after water injection is difficult to displaced by surfactant injection.
4. The presence of carbonates matrix or cement with Ca and Mg content cause solid precipitation during injection surfactant.

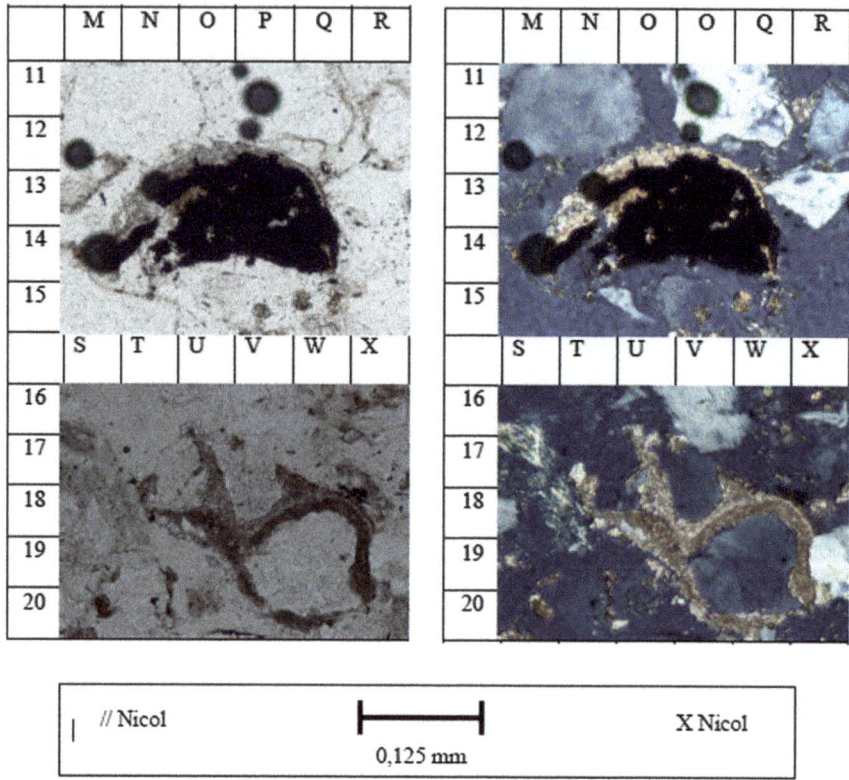

Fig. 8 Photomicrograph showingtrapped oil in the matrix and in the cavity of fossil shells, core sample of TPN #46

References

Barber, A.J., M.J. Crow, and J.S. Milsom. 2005. Sumatra geology, resources and tectonic evolution. *Geologycal Society Memoir*, 31.

Bishop, M.G. 2001. South sumatra basin province, Indonesia: The Lahat/Talang Akar-Cenozoic Total Petroleum System, U. S. *Geological Survey*.

Darman, H., and H. Sidi. 2000. An Outline of the Geology of Indonesia. *IAGI*, Chapter 2.

Grigg, R.B., and B. Bai. 2005. Sorption of Surfactant Used in CO_2 Flooding Onto Five Minerals and Three Porous Media. SPE 93100.

Marhaendrajana, T., and K. Idea. 2016. The effect of rock mineral and compositions on surfactant injection at tempino reservoir sandstone: A laboratory study, *JTMGB-IATMI*, 12:185–196.

Somasundaran P., M. Celic, A. Goyal, and E. Manev. 1984. The role of surfactant precipitation and redissolution in the adsorption of sulfonate on minerals. *SPE Journal*:233–239.

Webb N.D., J.P. Grube, C.S. Blakly, B. Seyler, and V. Madhavan. 2012. Reservoir Characterization of Lower Pennsylvanian Sandstones for the Application of ASP Flood Technology in Lawrence Field. Poster presentation at AAPG Annual Convention and Exhibition, Long Beach, California.

Williams, H., F.J. Turner, and C.M. Gilbert. 1982. Petrography: An introduction to the study of rocks in thin sections. 2nd Ed, W. H. Freeman, San Francisco.

Experimental Evaluation of Carbonated Water Injection to Increase Oil Recovery Using Spontaneous Imbibition

Enrico Adiputra, Leksono Mucharam and Silvya Dewi Rahmawati

Abstract Carbon dioxide flooding is known for increasing the production of oil as enhanced oil recovery (EOR). Conventional carbon dioxide flooding aims to reach minimum miscibility pressure (MMP) before altering oil properties in the reservoir. However, current conditions are that most fields have reached the mature state, so the reservoir pressure is depleted, thus it is hard to reach MMP. Carbonated water is water into which carbon dioxide has been dissolved, under certain conditions. The carbonated water injection (CWI) technique might be a solution for injecting carbon dioxide bellow MMP. The performance of this technique is examined using a physical model designed to show the wettability alteration mechanism of carbonated water. The physical model was made from a glass chamber filled with unaltered water. A saturated core was then placed inside the chamber below a funnel shaped narrow tube that read the oil recovery for each milliliter scale. The chamber was sealed and carbon dioxide injected into the water body. The water inside the chamber was therefore altered to become carbonated water after a period of soaking time. The oil expelled from the core was spontaneously measured by reading the scale on the top of the graduation tube, which showed how the milliliters oil was gathered, a process that known as imbibition. The process repeated for several concentrations of carbon dioxide in water. The change of injection pressure, power of hydrogen (pH), and oil recovery were measured respected to soaking time. The value of every case, including the unaltered-water-saturation case, were compared. This injection technique could result in 0–37% oil recovery.

Keywords Mature field · Carbonated water · Imbibition · Soaking time
Oil recovery

E. Adiputra (✉) · L. Mucharam · S. D. Rahmawati
Institut Teknologi Bandung, Bandung, Indonesia
e-mail: enricobpurba@gmail.com

© Springer Nature Singapore Pte Ltd. 2018
B. M. Negash et al. (eds.), *Selected Topics on Improved Oil Recovery*,
https://doi.org/10.1007/978-981-10-8450-8_4

Introduction

Production from mature fields accounts for more than 70% of current oil and gas production. Their recovery potential is enormous, with 80% of estimated reserves found in the Middle East and North Africa, 43% in Asia Pacific, and 24% in Latin America (About Mature Fields 2017). Several strategies involving secondary recovery, improved recovery, and enhanced recovery can be applied to increase the number. The challenge is finding a technology that can address the technical and economic aspects.

Carbonated water injection (CWI) can provide a solution to increase the recovery of oil without facing the difficulties of CO_2 injection to reach minimum miscibility pressure (MMP) (Lake 1989; Martin 1951; Sohrabi et al. 2011). In carbonated water, the CO_2 exists in a dissolved phase as opposed to a free phase, which eliminates the problems of gravity segregation and poor sweep efficiency on typical CO_2 injection projects (Sohrabi et al. 2011, 2015). CWI increases the recovery by several mechanisms: swell hydrocarbon, coalescence of trapped ganglia, and reduce viscosity (Sohrabi et al. 2015). Yet, there is another mechanism that can help to recover the hydrocarbon: the wettability alteration mechanism, which is anticipated to occur in CWI.

In the 1940s, Monteclaire Research from the Oil Recovery Corporation showed that S_{or} could be further reduced up to 15% PV if carbonated water was used after WF. Another experiment by Earlougher Engineering showed that residual oil saturation after carbonated water flooding was 2–26% PV less (Lake et al. 1984).

Johnson et al. reported that carbonated water flooding in sand packs at 75 °F and 750 psig carbonation pressure on low to medium viscous oil could increase the recovery by 15–25% after water flooding (Lake et al. 1984).

Martin reported a 12% improvement of oil recovery by using carbonated water which is better using the fully CO_2 saturated water instead of partially saturated water Martin (1951). This effect could be shown through a 14% reduction of high API oil and a 40% reduction of low API oil by dropping 50% carbonation. Sohrabi et al. 2015 used a micromodel to investigate *Carbonated Water Flooding* (CWF). They demonstrated that trapped ganglia could be recovered, resulting in 8.8% HCPV increase in recovery for light oil and 23.8% HCPV in viscous oil.

CWI has some mechanisms (Sohrabi et al. 2015) that may contribute to the improvement of oil recovery at various degrees, which are:

- Oil viscosity reduction
- Increase in water viscosity
- Oil swelling (increase saturation and the relative permeability)
- Wettability alteration
- Oil/water IFT reduction
- Injectivity improvement.

The objective of this paper is to investigate the CWI mechanism for enhanced oil recovery, especially the wettability alteration mechanism. The reaction of injecting

CO_2 into water (or so-called carbonation) along with the imbibition mechanism is explained in Chap. 2. The experiment procedures is presented in Chap. 3. The experiment results showing the applicability of CWI for increasing oil recovery is discussed in the Chap. 4. And, the Mole conclusion along with recommendations are discussed in the Chap. 5.

Carbonation and Imbibition

Carbonated water (H_2CO_3) could be made by injecting CO_2 into water under certain conditions to create a metastable reaction with high reversibility, but it needs a barrier or catalyst to be formed. In this case, pressure and temperature are maintained at a certain condition. The reaction between CO_2 and water follows the equilibrium reaction (1) and (2). The impact of temperature change to CO_2 solubility in water is shown in Fig. 1.

The process of imbibition, which is the process of absorbing wetting phase into porous rock (Amott 1959; Wael et al. 2007) is expected occur when soaking the

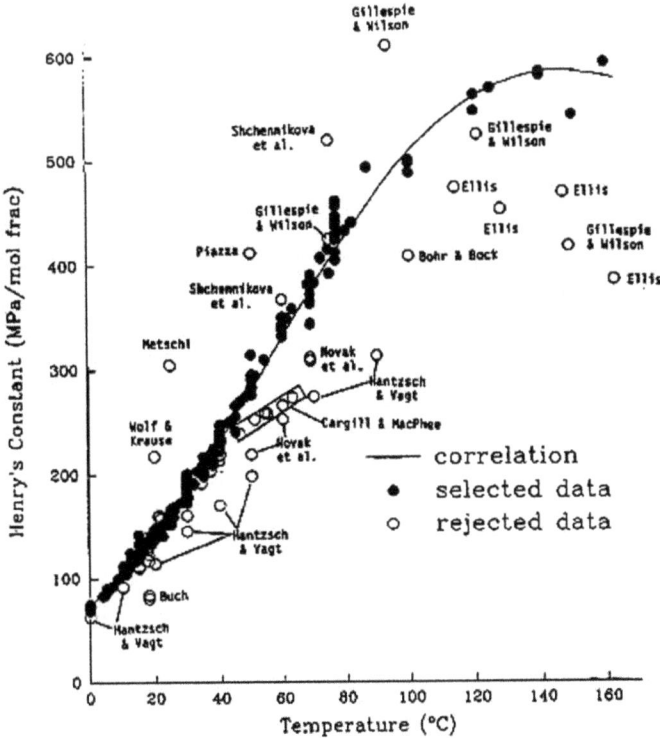

Fig. 1 Henry's constant for carbon dioxide in water—Carroll and Mather (1992)

reservoir with carbonated water. The Amott test method combines two spontaneous imbibition measurements and two forced displacement measurements could be used to measure the wettability alteration in most cases, however, this method could not be used as the pressure could not be maintained by a conventional Amott Cell. Therefore, a new physical model was made following a similar principle.

Experimental Setup and Procedures

Physical Model

The physical model was made with a glass vessel modified with a bonnet cap that bolted with a circular holder. Material for the vessel itself can hold up to 50 Psi injection pressure, however, due to the safety consideration, 1 bar of pressure was taken as a constrain. There are two valves and a pressure gauge on the cap of the vessel. The first valve (long pipe) designated to inject CO_2 into the water body. The second valve (short pipe) addressed as a vacuum and bleed for the pressure valve. Inside of the vessel, there was funnel shape narrowed tube and a core below it. The tube acted as a burette to measure the oil increase in the imbibition processes. The schematic for physical model is shown in Fig. 2.

Fig. 2 Physical model made to show the performance of CWI

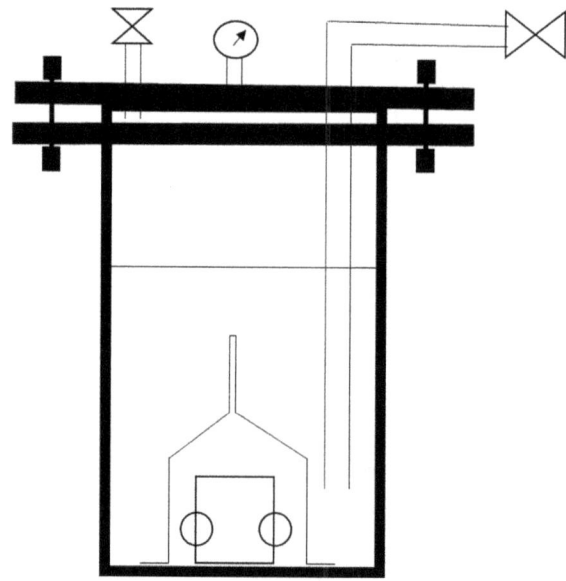

Table 1 Core properties

Core	Weight (g)		Volume (cm³)		Porosity
	Dry	Wet	Bulk	Pore	Fraction
RC-1 (B)	29.91	32.21	15.61	2.30	0.15
RC-2 (B)	30.43	32.48	15.79	2.05	0.13
RC-3 (B)	30.23	32.33	15.52	2.10	0.14
RC-4 (B)	31	33.01	15.79	2.01	0.13
RC-5 (A)	25.86	29.14	15.25	3.28	0.22
RC-6 (A)	24.79	28.14	14.45	3.35	0.23
RC-7 (A)	21.49	24.2	12.50	2.71	0.22
RC-8 (A)	22.75	25.82	12.86	3.07	0.24
RC-9 (A)	21.09	24.42	12.41	3.33	0.27

Table 2 Fluid properties

Name	Density	
	g/cm³	API
RO-1	0.81	42.25
Tap water	1	10.00

Artificial Cores

Two types of artificial cores with different physical properties were made. Type A was made with 50% cement and 50% quartz sand, which was calculated for 0.22–0.27 porosity using core saturation method. Yet, 40% cement and 60% quartz resulted in 0.13–0.15 fraction of porosity for Type B. The physical properties for each core can be seen in Table 1.

Crude Oil

One type of crude oil is measured using the pycnometer to find the mass for each volume of fluid. The properties of crude RO along with tap water as the denominator is shown in Table 2.

Imbibition Process

The imbibition process itself was separated into two parts: the water imbibition and the CWI imbibition. The water imbibition was done to find basic line before CWI. The recovery was measured by reading the oil volume that filled into the graduated tube. In this study, the measurement of recovery was done at 3, 8, 16, 24, and 48 h. The time considered the diminishing return effect, which means there will be less recovery overtime. The experiment temperature was at room temperature. The carbonation process was done after the water imbibition process was complete. The CO_2 gas was injected through the long pipe valve on the physical model. During the injection process, or the so-called carbonation process, bubbles were produced from the mouth of the pipe. The pressure gauge did not appear to be increasing several times as the CO_2 was moving from the bottom of the water to the free room above the water surface. After some time, the bubble production started decreasing and the pressure gauge increasing. After the desired pressure was reached, the injection process was terminated and the physical model closed completely to start the soaking period. The recovery was measured at 3, 8, 16, 24, and 48 h. The process repeated for 0.2, 0.5, 0.8, and 1 bar carbonation pressure.

Dissolved concentration of CO_2 in water could be estimated by Henry's Constant. The room temperature of 25 °C was used in this study, resulting 173.33 Mpa/mole fraction as the Henry's Constant. The mole fraction was found by dividing the carbonation pressure and Henry's Constant (Carroll and Mather 1992). Each concentration of CO_2 would yield a different value of pH, which was calculated from the first ionization process that produced H_3O^+. The pH was also measured using pH paper placed inside the physical model. The concentration for each carbonation pressure along with calculated and measured pH is shown in Table 3.

Result and Discussion

Spontaneous Imbibition

The water imbibition process yielded no recovery. Small droplets of oil that was shown on the surface of the core, however, the droplets were not moving upwards until toward the end of the experiment. The temperature might also have affected the recovery. Perhaps oil cannot be recovered on room temperature at given time.

The CWI yielded some recovery of oil sufficient to be read by the graduated tube. Overall, the RF was ranged from 0 to 37% of pore volume. The relation between RF and soaking time is shown in Fig. 3. The relation of RF and soaking time for each concentration is shown in Figs. 4, 5, 6 and 7.

On the 0.2 bar carbonation pressure, or calculated as 0.009 mol of CO_2, the recovery ranged between 0 and 11% of pore volume. At the first 3 h, no recovery was measured on the tube. However, some flocculation of oil was seen on the surface of the core, meaning the imbibition process occurred. On the 8 h state, the recovery could be seen by several experiments. The recovery increased after some time. However, the experiment by core RC-5 yielded a stagnancy that means there was no additional recovery after some point. This might have occurred because of leaks of

Table 3 Concentration for each carbonation pressure

Pressure (bara)	Mole fract	Water mole	Mole CO_2	Calculated			Measured		
				Water volume (L)	M CO_2	pH	pH RC-7	pH RC-8	pH RC-9
0.2	0.000115	77.713	0.009	1.4	0.006	4.28	5.50	3.50	4.50
0.5	0.000288	77.713	0.022	1.4	0.016	4.08	4.50	3.50	3.50
0.8	0.000462	77.713	0.036	1.4	0.026	3.98	4.50	3.50	3.50
1	0.000577	77.713	0.045	1.4	0.032	3.93	4.50	3.50	3.50

Fig. 3 RF versus time summary

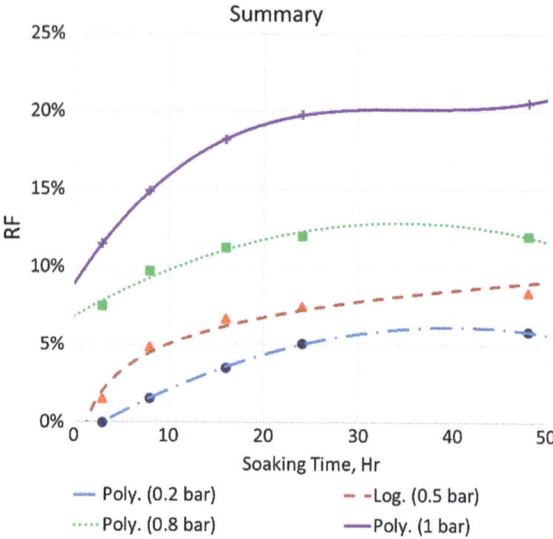

Fig. 4 RF versus time for 0.2 bar

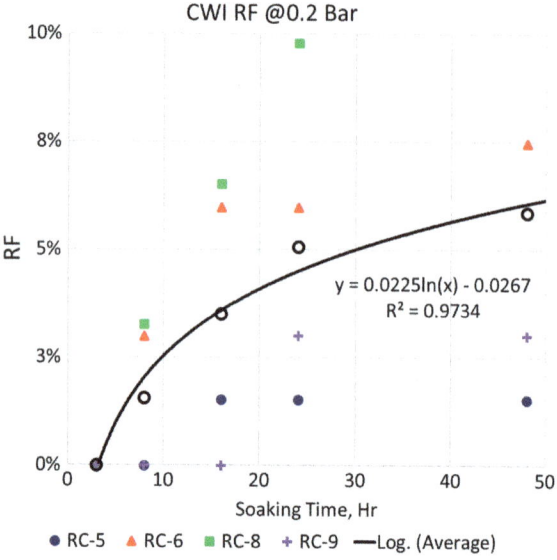

the physical model. Yet, the core RC-8 yielded the contrary result. The experiment on core RC-8 gave much more recovery as the time was increased. This occurred as RC-8 had greatly higher porosity that might also has good permeability. Overall, the diminishing return effect had not been seen in this concentration. There would still be an increase of recovery by time.

On the 0.5 bar carbonation pressure, or calculated as 0.022 mol of CO_2, the recovery ranged between 0 and 9% of pore volume. Some flocculation could be seen

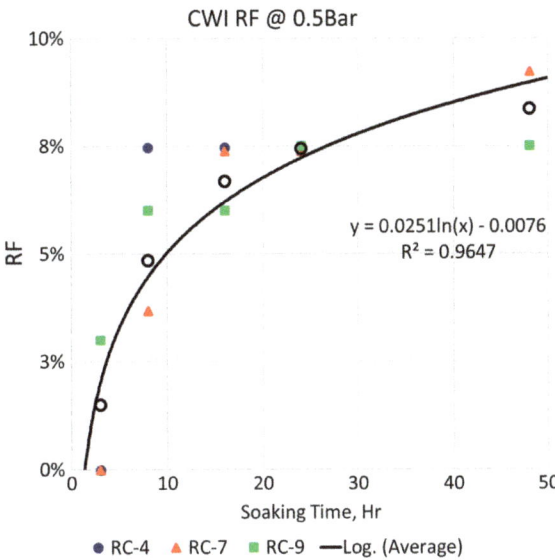

Fig. 5 RF versus time for 0.5 bar

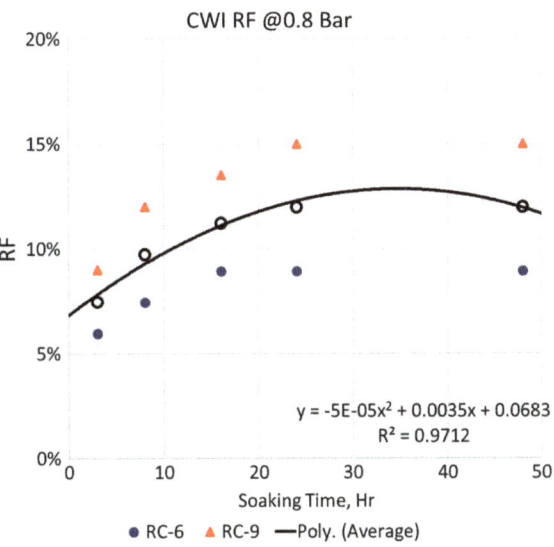

Fig. 6 RF versus time for 0.8 bar

with the smaller size than in the previous cases. This meant that better imbibition occurred in this concentration. The stagnancy also occurred in the experiment by RC-4 and RC-7 in this concentration. The stagnancy in this case might be taken as a leaks effect because the overall graph was not showing the diminishing return.

On the 0.8 bar carbonation pressure, or calculated as 0.036 mol of CO_2, the recovery ranged between 6 and 15% of pore volume. Some flocculation could be seen with an even smaller size than in previous cases. In this case, the diminishing

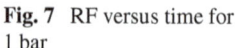

Fig. 7 RF versus time for
1 bar

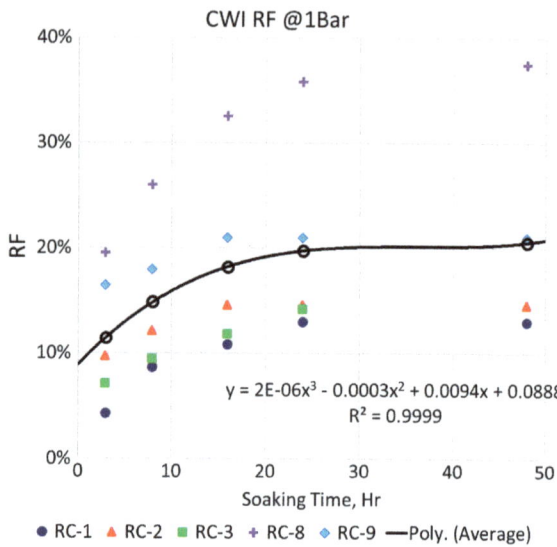

return effect could be seen. The optimum soaking time period in this concentration appeared at 35 h of soaking time.

On the 1 bar injection pressure, or calculated as 0.045 mol of CO_2, the recovery ranged between 4 and 37% of pore volume. This experiment comprised both sets of cores (Core A and Core B). The Core A sets involved RC-8 and RC-9. The Core B sets involved RC-1, RC-2, and RC-3. It was shown that the Core A sets were giving higher recovery as they had more porosity. The flocculation that even smaller than previous concentrations still occurred in some experiments. The diminishing return effect was clearly shown in this concentration with the optimum soaking time of 30 h.

The soaking time could not be estimated for 0.2 bar and 0.5 bar carbonation pressure cases as they had not reached the maximum recovery. However, a decision was made to terminate the soaking on those concentrations as a practical matter it was inefficient to wait through those long periods. It can be concluded that the bigger the injection pressure, the faster it reaches its maximum recovery. On the other hand, as is shown in Figs. 8 and 9, a linear curve occurred showing the relation between RF and concentration, which means the increase in concentration was still increasing during the recovery.

Field Application

The carbonation process happens on the surface, which means that CO_2 enriched water is injected into the reservoir as a liquid phase. The surface facility used in the CWI project should be easier to be maintained. There is no need for compressors

Fig. 8 RF versus CWI
concentration (pressure unit)
at 3 h

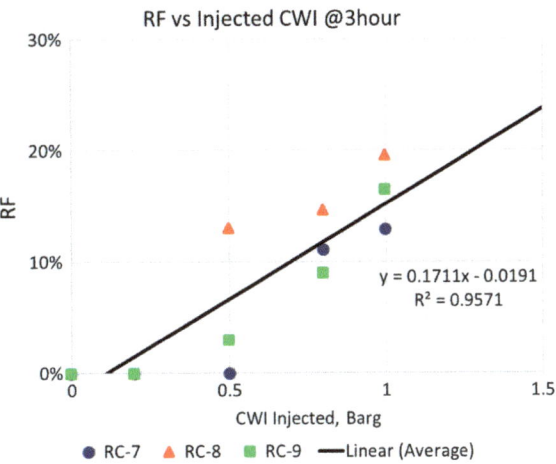

Fig. 9 RF versus CWI
concentration (pressure unit)
at 48 h

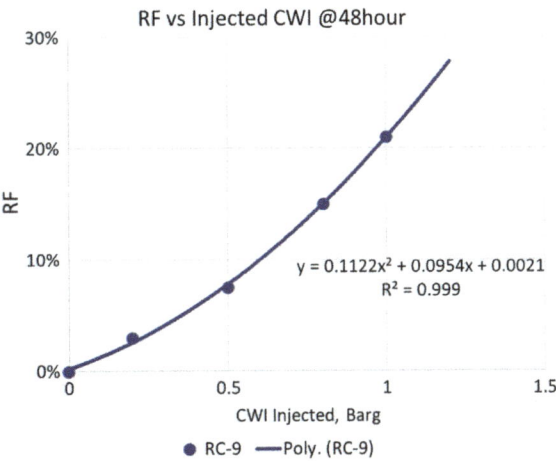

and there were only slight modifications to the current water flooding facilities. It was also beneficial that the CO_2 is easier to separate in the production facility as it is not dissolved in oil.

As a process that involves acid solution, the CWI could not be separated from corrosion problem. The corrosion was observed on the metal component of the physical model. This problem would be an issue if carbonated water was used under real conditions as the tubing itself is made from corroded material. More investigation into this issue is needed. A good simulation and calculation should be done to predict and solve this problem.

This experiment showed that a small little concentration of CO_2 could yield significant improvements to the recovery. Further, it demonstrated that carbonated water can be used in the reservoir with low pressure that far from MMP. In the end, carbonated water can become an alternative to conventional CO_2 injection.

Conclusion and Recommendation

Based on the experiments done in this study, physical model is designed to show the performance of CWI and the imbibition process by carbonated water is shown to be happened in this experiment. The recovery of the CWI ranged from 4 to 37% (optimum condition) and the recovery increase as the concentration is increased. The soaking time of CWI of is 30 h (optimum condition) and the optimum condition for this experiments is by 0.045 mol concentration. However, experiments using carbonate core, different type of crude oil, and more CO_2 concentration should be done to enrich the experience of CWI.

Attachment

$$CO_2 + H_2O \rightleftharpoons H_2CO_3 \tag{1}$$

or,

$$H_2CO_3 + H_2O \rightleftharpoons HCO_3^- + H_3O^+. \tag{2}$$

References

About Mature Fields. 2017. Halliburton. http://www.halliburton.com/en-US/ps/solutions/mature-fields/about-mature-fields. Accessed Oct 2017.

Amott, E. 1959. Observations Relating To The Wettability Of Porous Rock. *Transactions of the AIME* 219: 156–162.

Carroll, John J., and Alan E. Mather. 1992. The System Carbon Dioxide-Water and the Krichevsky-Kasarnovsky Equation. *Journal of Solution Chemistry* 21: 607–621.

Lake, L.W. 1989. *Enhanced Oil Recovery*. Published by Prentice-Hall.

Lake, L.W., G.A. Pope, G.F. Carey, and K. Sepehernoori. 1984. Isothermal, Multiphase, Multicomponent Fluid-Flow in Permeable Media. *In Situ* 8: 1–40.

Martin, J.W. 1951. Additional Oil Production Through Flooding with Carbonated Water. *Producer Monthly* 18–22.

Sohrabi, M., et al. 2011. *Carbonated Water Injection—A Productive Way of Using CO2 for Oil Recovery and CO2 Storage*. Elsevier Scientific Publishing Company.

Sohrabi, M., A. Emadi, S.A. Farzaneh, and Ireland, S. 2015. A Throughout Investigation of Mechanisms of Enhanced Oil Recovery by Carbonated Water Injection. Paper Presented at SPE Annual Technical Conference and Exhibition. Houston, Texas, U.S.A.

Wael Abdallah, Jill S. Buckley, Andrew Carnegie, John Edwards, Bernd Herold, dmund Fordham, Arne Graue, Tarek Habashy, Nikita Seleznev, Claude Signer, Hassan Hussain, Bernard Montaron, Murtaza Ziauddin. 2007. Fundamentals of Wettability. *Oilfield Review* 44–61.

Techno Economic Optimization of Hollow Fiber Membrane Design for CO$_2$ Separation Using Killer Whale Algorithm

Totok R. Biyanto, Andan Tanjung, Dimas B. Priantama, Tita Oxa Anggrea, Gabriella P. Dienanta, Titania N. Bethiana and Sonny Irawan

Abstract Natural gas has one of the main energy sources in the world. In the cryogenic processes of natural gas, the present of carbon dioxide remains a concern because it can cause the blockage of equipment. Membrane technologies are widely used in natural gas separation processes because of their compact size, simple operating conditions, no chemical additive. The type of membrane that used in this research is hollow fiber membrane type. The membrane optimization is aimed to obtain the best performance of membrane in term of technical and economical point of view by determine optimum membrane length that provide adequate residence time. Optimization of design hollow fiber membrane requires three components i.e. problem formulation, model and optimization techniques. The problem formulation of this research is to obtain the efficient designed hollow fiber membrane that provide maximum revenue, minimum cost and fouling, smaller size, and minimum cleaning. Modeling of dissolution and diffusion were built by utilizing Fick's law method, meanwhile pressure drop of hollow fiber membrane was modeled by Darcy equation. Optimization techniques were used in this research is Killer Whale Algorithm. Optimization results using Killer Whale Algorithm with parameters 20 matrilines, 5 leaders and 20 iterations were obtained the length of membrane 4 m, cleaning time interval 664 min, price of membrane 25,533,000 IDR/m^2 and the optimum revenue 1,162,000,000 IDR. Based on this research, the proposed techno economy optimization method can be applied in other the applications to solve energy problem.

Keywords Optimization · Hollow fiber membrane · Killer whale algorithm
Techno-economy

T. R. Biyanto (✉) · A. Tanjung · D. B. Priantama · T. O. Anggrea · G. P. Dienanta
Department of Physics Engineering, Sepuluh Nopember Institute of Technology (ITS), Surabaya, Indonesia
e-mail: trbiyanto@gmail.com

T. N. Bethiana
Department of Chemical Engineering, Sepuluh Nopember Institute of Technology (ITS), Surabaya, Indonesia

S. Irawan
Department of Petroleum Engineering, Universiti Teknologi PETRONAS, Seri Iskandar, Malaysia

© Springer Nature Singapore Pte Ltd. 2018
B. M. Negash et al. (eds.), *Selected Topics on Improved Oil Recovery*,
https://doi.org/10.1007/978-981-10-8450-8_5

Introduction

Natural gas still being the common option as the fuels in the world, moreover in Indonesia which has proven having 108.4 TCF (Trillion Cubic Feet) gas reserves as reported BP statistical review in world energy, 2011 (Pipkin et al. 2004). But before the natural gas that is brought from underground up to the wellhead to be a ready-used as energy, the natural gas needs to be pre-processed to remove contaminant such as CO_2, H_2S which constitutes environmental hazards and also cause hindrance in gas processing (Katcha 2010). H_2S tend to corrosive and CO_2 will reduce the thermal efficiency. CO_2 content in the typical pipeline specification is 2–5% before transported (Visser 2007). Hence it required to separation CO_2 gas.

There are several alternatives to CO_2 separation from natural gas by using physical separation (membrane and cryogenic), absorption (amine, hot potassium carbonate, flour, selexol and rectisol), and adsorption (Shamsabadi 2012). Membrane technology is being widely used, because it has advantages such as low cost, high reliability, easy to use, has a high on-stream operation, can eliminate high hydrocarbons, cheap maintenance, not really large in size, environmental friendly, and not using additives materials. In contrast to the filter which separating small molecules by porous medium, the membrane works by the diffusion principle of how well a compound dissolves and diffuses inside the membrane. At first the CO_2 dissolves into the membrane, then diffuses. The gas that can diffuse rapidly inside the membrane is called fast gas, e.g. CO_2, H_2, He, H_2S, and water vapor. Oppositely, the slow gasses are like CO, N_2, methane, ethane, and other hydrocarbons. Membranes are used to separate fast gas from slow gas (Shamsabadi 2012). The membrane consists of several types, including the hollow fiber membrane and the spiral wound membrane (Geankoplis 2003). This research uses hollow fiber membrane type because it has high reliability, easy to be operated, and the plan is generally smaller when compared with spiral wound membrane.

Based on research conducted by Dortmundt and Kishore Doshi (1999) the advantages of membrane technology compared to traditional CO_2 removal technology are smaller unit size, does not require cryogenic temperature in operation, simple utility requirement, high reliability, and easy to be operated (Dortmundt and Kishore Doshi 1999).

In this paper aim to obtain in terms of economy and design. Therefore, the application of optimization techniques is used. Basically the main problem which becomes the background of this paper is how to obtain the best membrane design and appropriate to get the appropriate output of natural gas containing residual little so that can be obtained minimum cost and maximum profit.

Optimization techniques require mathematical modeling and optimization methods in their use. The mathematical model used in this paper is based on Fick's law and Darcy's law. Meanwhile, the optimization method used is the method of Killer Whale Algorithm (KWA). KWA is the new algorithm efficient optimization technique in terms of function evaluation.

Fig. 1 Schema of hollow fiber

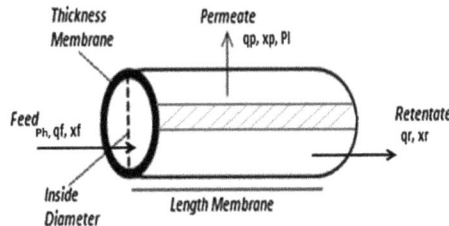

The membrane is often fouling due to contaminants which accumulate inside and on the surface of the membrane pore within a certain time. Fouling can be minimized by applying a backwashing system (Wu 2008). Fouling on the membrane may decrease permeability and increase operational costs.

Therefore, in this paper optimization will be presented on the design of hallow fiber membrane as the separation of carbon dioxide and methane. It aims to optimize membrane design so that low membrane price and optimum gain are obtained.

Methodology Research

Process Modelling

The membrane system is generally shown in Fig. 1 which is the simplified fiber hollow module with one fiber under cross flow operating conditions. The modeling shows that the feed flows along the outside of the hollow fiber. Then the gas is absorbed through the membrane wall and into the hollow fiber.

In accordance with the figure above, here is the following equation used to know the Material balance on the membrane:

$$qf \cdot xf = qr \cdot xr + qp \cdot xp \tag{1}$$

The equation above shows the equilibrium of the composition between the three gas fluids i.e. feed flow rate (qf), permeate flow rate (qp) and retentate flow rate (qr).

The area of membrane can be known using equation of tube area in general like Eq. 2 below. Membrane area is an influential factor to determine the parameter of ideal hollow fiber membrane design.

$$Am = \pi \cdot d \cdot L \tag{2}$$

where:

Am   Surface area (m^2)
L   Length of membran (m)

The amount of pressure drop on the permeate side can be found by equation based on Darcy's law.

$$\Delta P = \frac{qp \cdot \mu \cdot \ell}{K'A \cdot A} \tag{3}$$

where:

ΔP Pressure drop (Pa)
μ Permeate viscosity (Pa min)
A   Cross sectional area (m^2)
K   Permeate permeability (m^3/min m^2 Pa)

To know the magnitude of the pump work (W_p) then used the following equation:

$$W_p = \frac{\Delta P \times Q}{\eta} \tag{4}$$

And, the cost of pump use (C_{pump}) is calculated in the following way:

$$C_{pump} = W_p \times electricity\ cost \tag{5}$$

To know the cost of production, we can use the following equation:

$$C_{pro} = qf \cdot C_{ch4} \tag{6}$$

where:

C_{pro} Cost of production (IDR/m^3)
C_{ch4} Cost of methane (IDR/m^3).

Determination of Objective Function

The purpose of this method is to determine the objective function used for optimization on the design of this membrane. The following is the objective function equation used.

$$J_{max} = (-(Cm \times Am) - ((W_p \times Cl \times 10) + (Y \times C_{pro}))) \tag{7}$$

where:

Cm Cost of membrane (IDR/m^2)
Cl   Cost of electricity (IDR/kW $\times$ 10 min)
t   Backwashing time (10 min)
Y   Status of backwashing (0 when off and 1 when on process).

Tabel 1 Parameters of KWA

Parameters	Value
Matrilines	20
Variable	2
Leader	5
Member	15
Maksimum iteration	20

In this objective function, the membrane price refers to a study conducted by Richard and Kaaeid for $500/m² (Richard 2007). As for the cost of electricity per kWh refers to the price of electricity in October 2016 PLN 1032.62 IDR/kWh.

Fouling Modeling

The problem which often encountered in the membrane process is the tendency for flux to decrease during operation time due to precipitation or adhesion of material on the membrane surface or known as fouling.

Fouling can cause a higher workload, thus causing an increase in energy consumption which affects the declining economic value of the system. Based on these economic considerations, then there is urgent need to reduce the fouling can be done regular cleaning.

The fouling modeling can be done by regression method. This regression method is usually used for clearance scheduling. Over time, the membrane permeability will decrease. This is due to the formation of fouling. Here is the equation of the regression method:

$$y = 0.00002 - 0.0000000005t \tag{8}$$

where:

y   Permeability (m³/min m² Pa)
t   Time (min).

Killer Whale Algorithm

Killer Whale Algorithm in this research is designed to be able to produce optimum value. In this optimization, the objective function utilized to obtain minimum operational cost with maximum profit. In addition to the objective function, some parameters on KWA should also be considered Table 1.

Table 2 Input parameter of carbon dioxide separation process from methane

Parameter	Value	Unit
Feed flow rate	0.6	m^3/minutes
Fiber diameter	0.304	m
Viscosity	0.22833	Pa minutes
Pum efficiency	0.8	–
Membrane thickness	0.03	m
Initial pressure	250,000	Pa

The initialization process is performed to determine the initial parameters such as, the number of Killer Whale populations, the dimensions of the objective function, the global optimum condition between finding the minimum or maximum value, the lower limit and the upper limit of the objective function, the number of clusters as well as the number of iterations for the clustering process (Biyanto et al. 2017).

The cluster method is utilized to accelerate the search for global optimum values in objective functions and to avoid the final result of the local optimum value, the population will be divided into two, between Leaders and Members. The search area in each cluster will be tracked by both groups, the search process begins at the center point of the mass of each cluster, the Leader will move to the other side of the cluster if the value obtained is lower than the Member. If the Leader switches to another cluster, then the Member will also move to another cluster, this process is done using the principle of the number of clusters that divide the number of iterations as there are 4 clusters with 20 iterations for the search agent, then to one cluster doing 5 iterations (Biyanto 2017) (Fig. 2).

Results and Discussions

Modeling of Carbon Dioxide Separation Process from Methane

This research begins by checking the design data using material balance. The principle of this step is to equalize the flow rate given to the input by the amount of flow rate on the permeate side and on the retentate side. After calculating the material balance, the next step is to calculate the pressure drop to know the cost of compressor. The cost of this compressor also affects operational costs. The parameters conditions and input operations used in the calculations should also be noticed Table 2.

After the calculation of material balance, then the next step is to calculate the membrane length, flux, flow rate, pressure drop and operational cost. This calculation processes use Eqs. (1)–(7). From the calculations then we can obtained that the flux decreases by time Fig. 3 while the pressure drop increases by time Fig. 4.

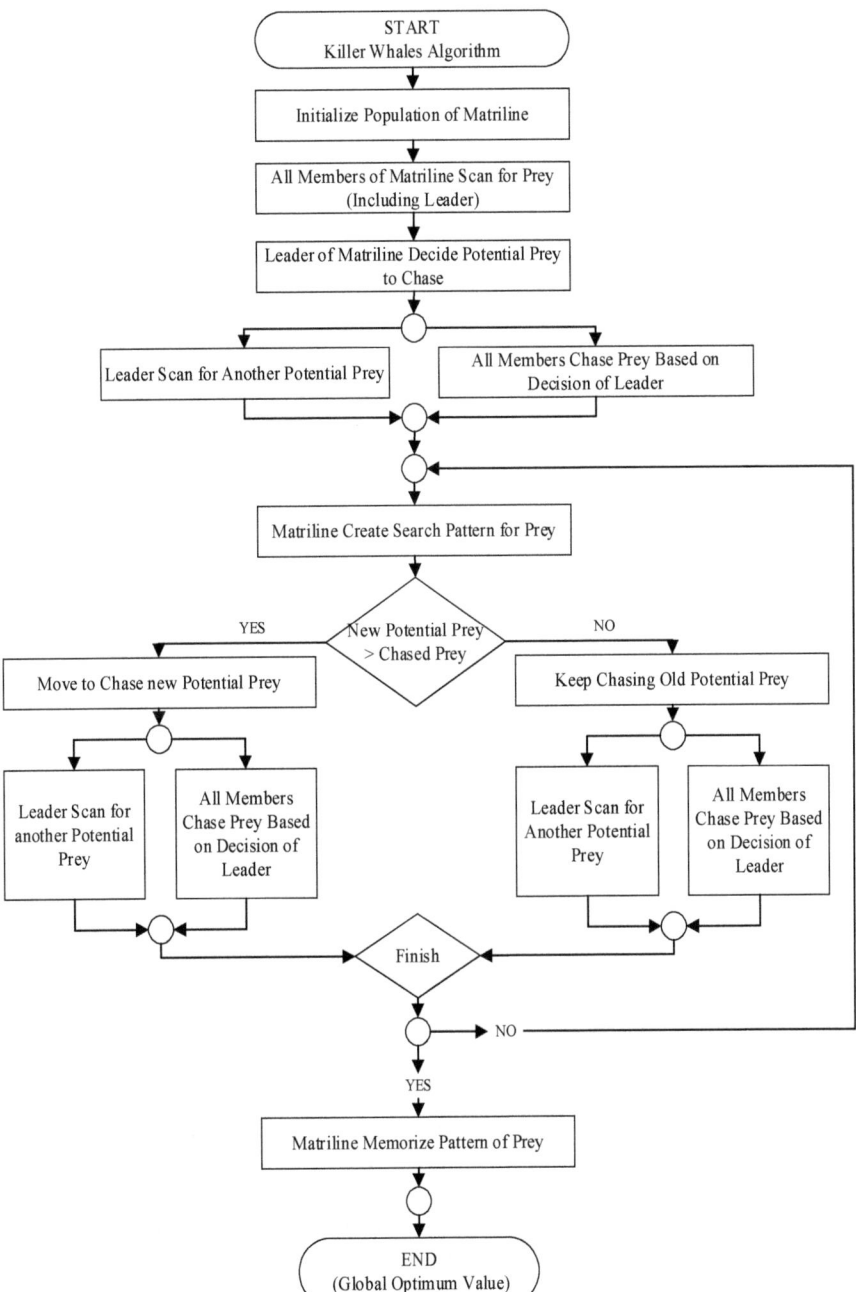

Fig. 2 Flow chart of killer whale algorithm

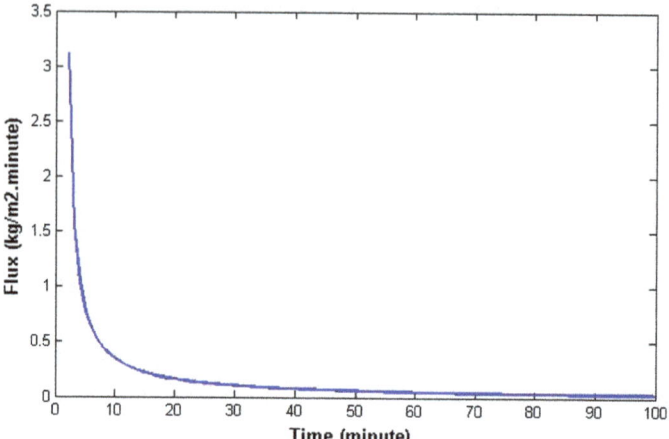

Fig. 3 Decreasing of flux by time

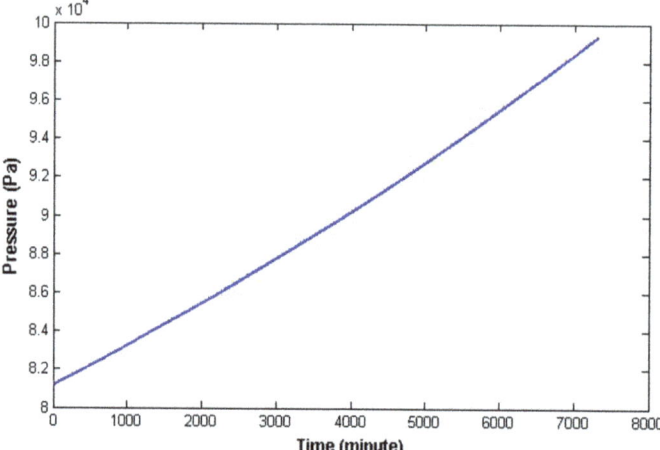

Fig. 4 Increasing of pressure drop

During the process of separating carbon dioxide from methane, membrane performance may change or decrease by increasing the time. As shown in Fig. 3 which shows a decrease in flux as time increases. This indicates that the membrane pores begin to be covered by the presence of a precipitate. In the figure, the graph shows the decrease of flux in the hollow fiber membrane up to the number 0.0316 (kg/m^2 min). Samples taken on this flux reduction chart are limited to 100 min duration. This is done to clearly show the value of flux reduction, according to the figure. Based on research that has been done, the parameter which decreases over time is not only flux, but also the membrane permeability. One other factor of flux reduction and permeability is the emergence of fouling. Foulant will accumulate on the

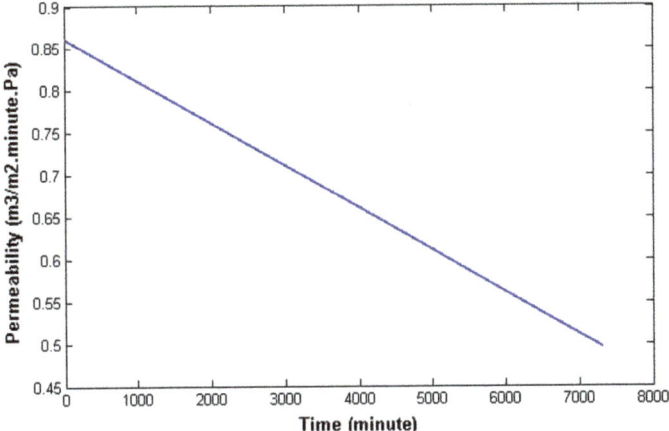

Fig. 5 Relation of permeability and time

surface of the membrane because it does not take part in mass transfer. As a result this foulant will reduce the effectiveness and flux of the membrane. Foulants can be organic sediments (macromolecules, biological substances), inorganic deposits (metal hydroxides, calcium salts) and particulates.

The Fig. 4 shows that pressure drop increasing by the time, the value difference of the pressure will affect the magnitude of the pressure on the permeate side and impact on the large compositions that penetrate the membrane. The increase in pressure is caused by the smaller permeability of the membrane due to the emergence of fouling.

Fouling

In addition to the flux, the membrane permeability will decrease by the time Fig. 5. This is due to the formation of fouling. Figure 5 shows a decrease in permeability. The decreasing of permeability shows that the membrane performance is not optimal. In order to improve the performance of a membrane, it is necessary to clean it periodically by performing a backwashing system on the membrane. Backwash duration greatly affects permeability. In accordance with the reference obtained, in this study the duration of backwashing done for 10 min.

Optimization

After the modeling, then the optimization is done with optimized variables is the length of membrane fiber and time interval Fig. 6. Optimization is done by using

Fig. 6 The maximize
objective function of fitness
during iterations

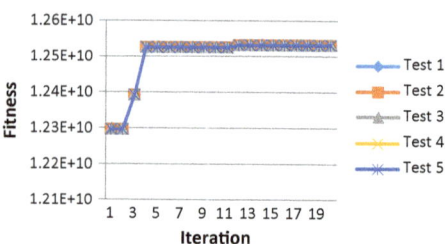

killer whale algorithm. The optimization of killer whale is performed to determine the value of membrane fiber length which has an effect on the membrane area and the operating cost with the objective function shown in Eq. (7). The value of the killer whale parameter used in the optimization is shown in Table 1. The optimization of this membrane design, the length and time interval values are obtained randomly, through optimization using killer whale. Where the optimization is used limits of length of 3–5 m, and for time interval has a limit of 0–7200 min. After the optimal value and time interval is obtained, calculation of pressure drop, flux, to operational cost and membrane price is calculated. Basically optimization on the plan is done to obtain the optimal length and time interval. Figure 6 is a graph showing optimization results using the killer whale algorithm method.

Optimization Results

The optimization result using killer whale algorithm, the optimum length of membrane is 4 m. This values have an effect on the operational cost Figs. 9, 10 and 11. In addition to membrane fiber length, optimization is also done on the time interval and the value obtained is 664 min. So it can be interpreted that each membrane operates for 664 min, it must be done backwashing system. This is necessary to reduce the appearance of fouling. The more often it is backwashed then the better is the reduction of the fouling.

After optimization, the membrane separation process with methane is designed with the length of fiber membrane 4 m and cleaned periodically then the maximum membrane price is 25,533,000 IDR/m² Fig. 7.

The results showed that with the application of the backwash system for 10 min, the magnitude of the membrane permeability rises to 0.86 (1/m² min Pa) Fig. 8. This indicates that only a small amount of CO_2 gas is trapped in the membrane pores that cannot set aside with backwashing. The increase in the value of a permeability indicates that the membrane performance increases from before. Thus, it affects pressure difference and the product of the separation of carbon dioxide gas with methane. The smaller the permeability, the higher the pressure difference and the smaller the carbon dioxide gas that penetrates the membrane. This is shown in Figs. 9, and 10 show the cost of pumping for 7200 min resulting in a maximum value of 2.293

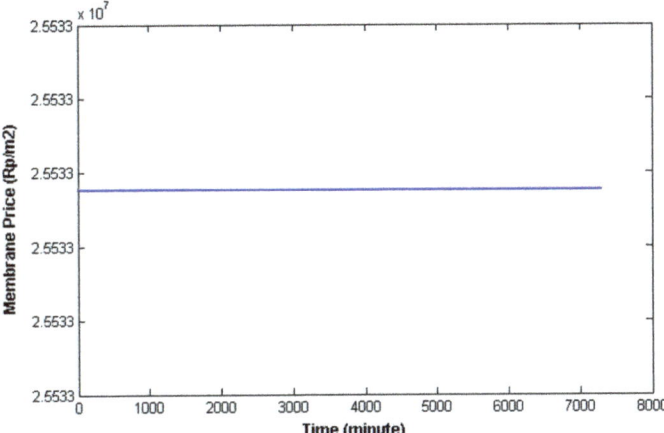

Fig. 7 Membrane cost

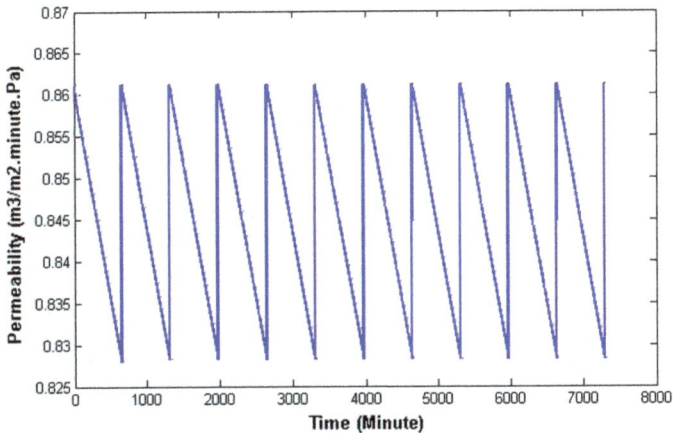

Fig. 8 Permeablity value after using backwash system

IDR/kW min. Meanwhile in Fig. 11, the result shows that the maximum value of production price is 1.136.000.000 IDR/m^3.

Thus, it can be seen that the longer the membrane operates, the performance decreases but the higher the pressure due to the emergence of fouling. The occurrence of fouling can be reduced by a backwash system which in this study is carried out with a duration of 10 min, and is applied every 664 min. Given such an application, the maximum profit gained according to the objective function is 1.162.000.000 IDR/month Fig. 12.

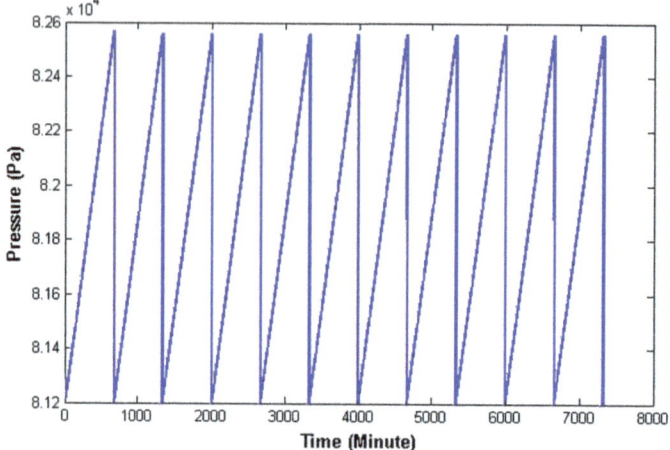

Fig. 9 Pressure difference after using backwash system

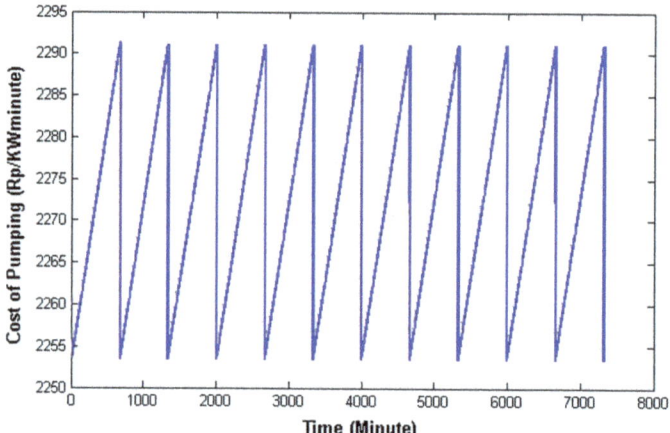

Fig. 10 Cost of pumping after using backwash system

Conclusions

The analysis of mathematical modeling results conducted using secondary data, it is known that there is a decrease of permeability due to fouling and increase of pressure drop. Based on the optimization result by using killer whale algorithm on carbon dioxide separation system with methane, the optimum membrane length is 4 meters, while for time interval is 664 min. Hence for the prevention of membrane fouling should be cleaned every 664 min. The application of optimization membrane prices obtained 25,533,000 IDR/m^2 and maximum pump usage fee of 2293 IDR/kW min and production cost of 1,136,000,000 IDR/m^3. Thus, in accordance with the objective

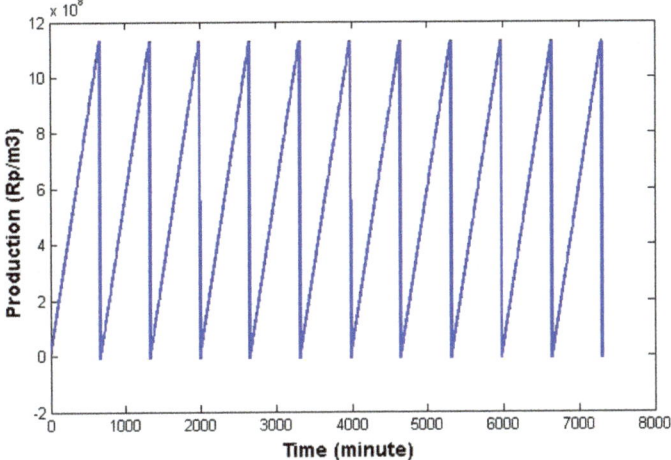

Fig. 11 Production cost after using backwash system

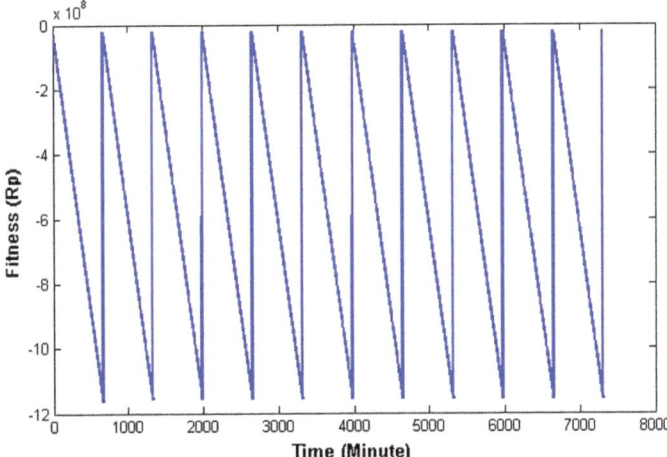

Fig. 12 Fitness function by time

function used, the maximum profit obtained 1,162,000,000 IDR in cleaning as much as 11 times.

Acknowledgements The author gratefully thank to Mr. Totok Ruki Biyanto formerly Sepuluh Nopember Institute of Technology lecturer for helping makes this research possible, and for his many important contributions to the development of Petroleum and energy sector.

References

Biyanto, Totok R., Matradji, Sawal, Ahmad Hasinur, Rahman, Arfiq Isa, Abdillah. 2017a. Application of Killer Whale Algorithm in ASP EOR Optimization. *Procedia Computer Science* 124: 158–166.

Biyanto, T.R., Matradji, Sonny, Irawan, Henokh Y. Febrianto, Naindar, Afdanny, and Ahmad Hasinur, Rahman, Kevin Sanjoyo, Gunawan, Januar A.D. Pratama. 2017b. Killer Whale Algorithm: An Algorithm Inspired by the life of Killer Whale. *Procedia Computer Science* 124: 151–157.

Dortmundt and Kishore Doshi (1999). Recent Developments in CO_2 Removal Membran Technology. *UOP LLC* 1–32.

Geankoplis, C.J. 2003. *Transport Process and Unit Operations*, 4th ed. New Jersey: Prentice Hall Professional Technical Reference.

Katcha, S. 2010. Membran Technologies For CO_2 Capture.

Pipkin, Bernard W., Dee D. Trent, Richard, Hazlett, and Paul Bierman. 2004. *Geology and the Environment*, 7th ed. Cengage Learning.

Richard, W. 2007. *Natural Gas Processing with Membranes*. California.

Shamsabadi, A.A., A. Kargari, F. Farshadpour, and S. Laki. 2012. Mathematical Modeling of CO_2/CH_4 Separation by Hollow Fiber Membran Module Using Finite Difference Method. *Journal of Membrane and Separation Technology* 19–29.

Visser, T., N. Masetto, and M. Wessling. 2007. Materials Dependence of Mixed Gas Plasticization Behavior in Asymmetric Membranes. *Journal of Membrane Science* 306: 16–28.

Wu, J., P. Le-Clech, R.M. Stuetz, A.G. Fane, and V. Chen. 2008. Effects of Relaxation and Backwashing Conditions on Fouling in Membran Bioreactor. *Journal of Membrane Science* 324: 26–32.

Sensitivity Analysis of CO_2 Injection to Oil Rate Using COZView Simulator

Leovaldo Pangaribuan and Ully Zakyatul Husna

Abstract In achieving the program goals of CO_2 injection, as one of enhanced oil recovery method, some parameters are needed to be highly considered. Sensitivity study is usually aim to evaluate the effect of varying parameters on the reservoir performance. In this work, CO_2 injection simulation was conducted using simulator named COZView. The main objective of this paper is to analyze sensitivity of parameters that affect CO_2 injection performance to oil rate. Those parameters are temperature, API gravity and injection rate. The results have showed that, after two years of production, the model is found to be most sensitive to temperature and API gravity. The higher API gravity, the greater oil rate. This relation have similar to temperature and injection rate. But, for the injection rate case, there is critical condition as depending of reservoir itself conditions. Thus, it is recommended that these data must be properly selected and analyze to optimization of CO_2 injection performance for the good results in the oil field.

Keywords CO_2 injection · Temperature · API gravity · Injection rate COZView

Introduction

The oil and gas industry continuously developing experience in injecting CO_2 for enhanced oil recovery (EOR) since early 1950s (Ifeanyichukwu et al. 2014). Field experience has shown that CO_2 injection is one of effective method to improve oil recovery (Yin 2015). For example, there are more than 100 CO_2 flooding projects producing more than 250,000 barrels of oil per day in US (Zhou et al. 2012). This CO_2 method improves oil recovery by lowering interfacial tension, swelling the oil, reducing oil viscosity and reducing oil density (Verma 2015).

L. Pangaribuan (✉) · U. Z. Husna
Department of Petroleum Engineering, Faculty of Engineering, Universitas Islam Riau, Jl. Kaharuddin Nasution No. 113 Km. 11, Pekanbaru, Riau 28284, Indonesia
e-mail: leovaldopangaribuan@student.uir.ac.id

© Springer Nature Singapore Pte Ltd. 2018
B. M. Negash et al. (eds.), *Selected Topics on Improved Oil Recovery*,
https://doi.org/10.1007/978-981-10-8450-8_6

Table 1 Reservoir properties of Permian Basin (Safi 2015)

Reservoir thickness	100 ft
Porosity	0.12
Reservoir temperature	235 °F
Initial pressure@−4500 ft ss	1500 psia
Initial bubble point pressure	800 psia
Horizontal permeability	150 Md
Vertical Permeability	15 mD
Boundary condition	No mass flow at boundaries
Reservoir volume	2,137,285 bbl
Oil gravity	40.7 API
Water salinity	70,000 ppm
HC gas specific gravity	0.7

However, in achieving the successful project of CO_2 injection, some properties are needed to be considered. Those parameters can be both the reservoir itself, or operation parameters. In Ifeanyichukwu et al. studies (2014), the results have shown that reservoir heterogeneity and dip, CO_2 to oil mobility ratio, injection rate, volume of CO_2 injected, reservoir fluid nature, and injection well configurations are important parameter that affect the performance of CO_2 EOR projects.

Besides, COZView Simulator is a tool that allows user to analyze optimization of CO_2 injection. It was developed by NITEC LLC under a Federal Assistance Agreement with the U.S. Department of Energy/National Energy Technology Laboratory (COZView/COZSim User Manual 2017). This simulator provides some results that show relation of production aspects to the well or field model. In this paper, the main purpose of this study is to analyze sensitivity parameters that affect to CO_2 injection performance. The parameter that chosen to be candidates of sensitivity are temperature reservoir, API gravity of crude oil, and injection rate.

Model Description

The conceptual model used in this paper using data from a real industrial scale reservoir in the Permian Basin. The model represents five-spot pattern with 4 producer wells and 1 injection well. The computational domain is discretized into a 3 × 3 grid. An injection well and producer wells are placed on opposite corners, which the wells are perforated in each of the 4 layers with a thickness of 25 ft for each other (Safi 2015).

The reservoir characteristics are summarized in Table 1.

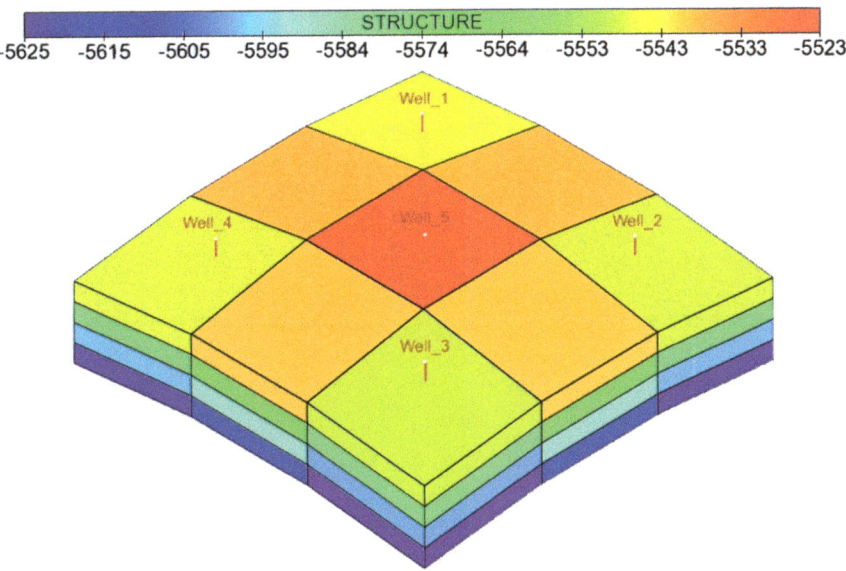

Fig. 1 Structure model (COZView/COZSim User Manual 2017; COZView/COZSim Tutorial 1 2017

Simulation Cases

The simulations for this study were implemented using black oil simulator that is capable of modeling CO$_2$ injection with vary parameters. This study was focused about sensitivity of oil PVT data and field constraint to oil rate (Fig. 1).

There are more than five cases of each sensitivity parameter. All the cases are continuous CO$_2$ injection for two years. The parameter that chosen to be candidates of sensitivity are temperature reservoir, API gravity of crude oil, and injection rate. These parameters will varying that are used to evaluate the effect to the oil rate performance. For the temperature, there are ten cases that is range from 104.27 to 300 °F to be runs, where five cases are the property will be decreased 15% of the actual value and the rest cases are increased 5% of the actual value. For the API gravity, there are eight cases that started from 18 API to 47.1 API to be runs, where four cases first are the property will be decreased 15% of the actual value and the other cases are increased 15% of the actual value. For the injection rate, there are seven cases to be runs where all cases are increase up to 5000 MSCFD.

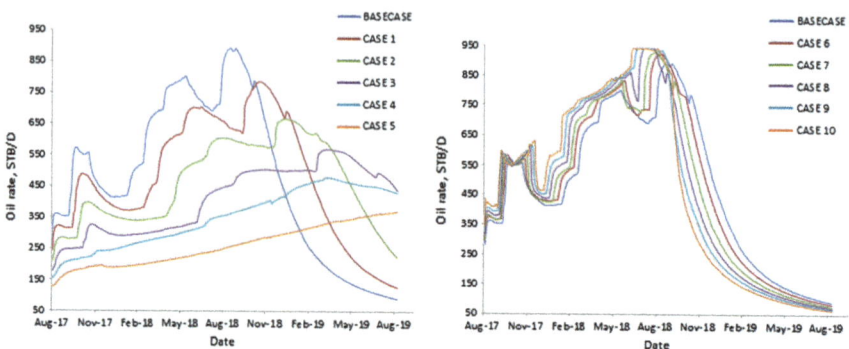

Fig. 2 Date versus oil rate (temperature)

Results & Discussion

Based on Fig. 2, the reservoir temperature affect to the oil rate performance. It can be seen that the higher temperature will result the increasing of oil rate, however the decreasing rate time is fast happened. These results have similar to the theory, where the temperature is related to viscosity. The viscosity is lower with higher temperature (Yin 2015). It has been implied by using Vasquez and Beggs correlation (Verma 2015), like temperature 235 °F have viscosity 0.91 cp and temperature 272 °F have viscosity 0.73 cp. Viscosity affects the mobility ratio of the fluids. If oil viscosity is small, then the mobility ratio will be small also. The lower the mobility ratio, the greater is the oil recovery (Nasir and Amiruddin 2008).

But, for focusing of CO_2 matter, increasing the temperature causes the kinetic energy of CO_2 molecules will increase also, which creates a more active CO_2 phase. This may cause a higher degree of contact of CO_2 molecules with residual oils in the reservoir, which increases oil production. It is probable that the effect of kinetic energy increment and the corresponding increment in CO_2 mobility with increasing temperature is much higher than the CO_2 vaporizing effect under high temperature conditions (>212 °F) (Perera 2016) (Table 2).

In these cases, CO_2 minimum miscibility pressure by using Yelling and Metcalfe correlation (Yuan 2005) were calculated between 1267 and 3783 psi. It means almost all cases are immiscible flooding. By this results, we also found that MMP increases linearly corresponding to temperature. MMP increases as temperature increases (Yin 2015).

Even miscibility cannot be reached for this some reservoir condition, a high recovery rate still can be achieved like these simulation results. It mainly due to oil swelling as it becomes saturated with CO_2, viscosity reduction of the swollen oil and CO_2 mixture, and reservoir size (Yin 2015).

In other case, theory and field applications both demonstrate that the light oil reservoir are better candidates for CO_2 flooding (Chukwudeme 2009). It is similar

Table 2 Cases of temperature

	Base case	Case 1	Case 2	Case 3	Case 4	Case 5	Case 6	Case 7	Case 8	Case 9	Case 10
T (°F)	235	199.75	169.79	144.32	122.67	104.27	246.75	259	272	285.6	300

from the result of Fig. 3. As high oil gravity, oil rate will be higher. It caused by CO_2 will mobilizing the lighter components of the oil.

However, as lower API gravity or heavier the crude oil, oil rate will be smaller. This is because heavy oil will tend to resist gas mobility (Chukwudeme 2009), and CO_2 lacks acceptable sweep efficiency due to the large viscosity contrast between CO_2 and oil as well as unlikely development of a miscible front in heavy oil reservoirs. Thus, results show that effect of CO_2 is not works optimally to improve oil recovery (Table 3).

For the injection rate cases, injection rate starts from 1500 MSCF up to 5000 MSCF. It is not considered that an optimal rate of CO_2 will be greater than 5000 MSCF/day, because it is not only unrealistic economically for this reservoir size matter, but also will cause pressure increase that may yields fracture in the reservoir (Safi 2015) (Table 4).

Based on Fig. 4 implies that increasing of injection rate cause oil production will increases. It makes pressure will be higher, so this high pressure will help to attain CO_2 miscibility pressure. However, in case above 2500 MSCF, it is not really affecting to the oil rate. This term also was supported by the result of recovery factor that described in Fig. 5, where injection rate above 3000 MSCF has similar results. It caused by higher injection will make time of CO_2 breakthrough fast occurs. When breakthrough occurs, gas injected will not help effective to oil rate, as the gas will goes faster towards the producer wells (Khurshid 2013). By this cases, optimum injection

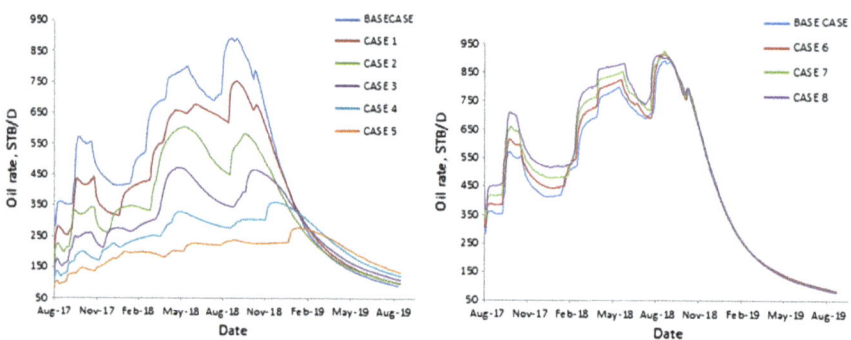

Fig. 3 Date versus oil rate (API gravity)

Table 3 Cases of API gravity

	Base case	Case 1	Case 2	Case 3	Case 4	Case 5	Case 6	Case 7	Case 8
API gravity	41	34.59	29.4	25	21.24	18	42.7	44.8	47.1

Table 4 Cases of Injection Rate

	Base case	Case 1	Case 2	Case 3	Case 4	Case 5	Case 6	Case 7
Injection rate, MSCF	1500	2000	2500	3000	3500	4000	4500	5000

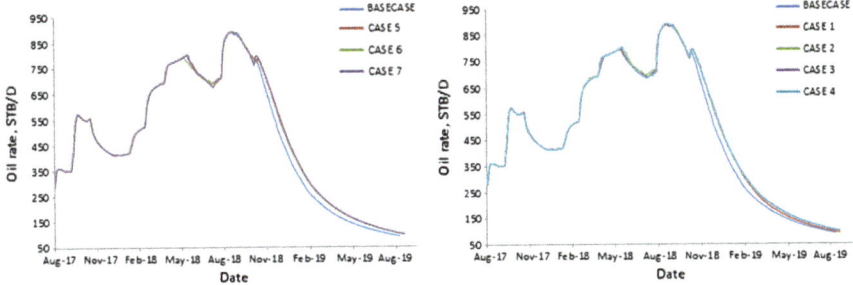

Fig. 4 Date versus oil rate (injection rate)

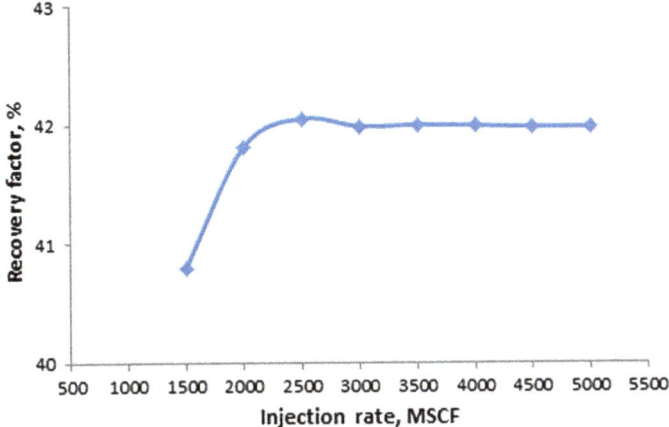

Fig. 5 Injection rate versus RF

rate for this reservoir condition is 3000 MSCF that have the biggest recovery factor 42.1%.

Conclusion

The main objective of this study is to analyze sensitivity of parameters that affect to CO_2 injection performance. Those parameter are temperature, API gravity and injection rate. The higher temperature will result the increasing of oil rate, however the decreasing rate time is fast happened. In other hand, as lower API gravity or heavier the crude oil, oil rate will be smaller. For injection rate cases, as high injection rate, oil rate will be increase, but there is critical condition as depending of reservoir itself.

Acknowledgements Authors would like to thank Department of Petroleum Engineering, Faculty of Engineering, Universitas Islam Riau for any support in writing this paper. Authors would like to thank Dr. Eng Muslim for his valuable guidance and enlightening in this study. Authors would also like to thank COZview Simulator for allows us to analyzing sensitivity of CO_2 injection performances to oil rate.

References

Chukwudeme, E.A, and A.A. Hamouda. 2009. Enhanced Oil Recovery (EOR) By Miscible CO_2 and Water Flooding of Asphaltenic and Non-Asphaltenic Oils. *Journal Energies*, 6.

COZView/COZSim User Manual. 2017. Nitec, LLC website, http://nitecllc.com/docs/COZView_UserManual.pdf, 20 August, 2017.

COZView/COZSim Tutorial 1. 2017. Nitec LLC website. http://nitecllc.com/docs/Tutorial 1.pdf, 20 August, 2017.

Ifeanyichukwu, P.C., J.U. Akpabio, and S.O. Isehunwa. 2014. Improved Oil Recovery by Carbon Dioxide Flooding. *International Journal of Engineering and Technology* 4 (5): 1–6.

Khurshid, I., and J. Choe. 2013. Characterizing Formation Damage Due To Carbon Dioxide Injection in High Temperature Reservoirs and Determining the Effect of Solid Precipitation and Permeability Reduction on Oil Production, SPE 165158, Presented at the SPE European Formation Damage Conference and Exhibition, Netherlands, June 5–7 2013.

Nasir, F.M., and N.A. Amiruddin. 2008. Miscible CO_2 Injection: Sensitivity to Fluid Properties, SPE 115314, Presented at the SPE Asia Pasific Oil & Gas Conference and Exhibition. *Perth, Australia* 20–22: 5–6.

Perera, M.S.A, R.P. Gamage, T.D. Rathnaweera, A.S. Ranathunga, A. Koay, and X.A. Choi. 2016. A Review of CO_2 Enhanced Oil Recovery with a Simulated Sensitivity Analysis. *Journal Energies* 9 (7): 481.

Safi, R. 2015. *Numerical Simulation and Optimization of Carbon Dioxide Utilization for Enhanced Oil Recovery from Depleted Reservoirs*, Master thesis, Washington University.

Verma, M.K. 2015. *Fundamental of Carbon Dioxide-Enhanced Oil Recovery (CO_2-EOR)—A Supporting Document of the Assessment Methodology for Hydrocarbon Recovery Using CO_2-EOR Associated with Carbon Sequestration*, Geological Survey File Report.

Yin, M. 2015. *CO_2 Miscible Flooding Application and Screening Criteria*, Master thesis, Missouri University of Science and Technology.

Yuan, H, R.T. Johns, Egwuenu, and B. Dindoruk. 2005. Improved MMP Correlations for CO_2 Floods Using Analytical Gasflooding Theory, SPE 89359, Presented at the SPE Symposium on Improved Oil Recovery, Tulsa, April 17–21 2005:7.

Zhou, D., M. Yan, and M.C. Calvin. 2012. Optimization of a Mature CO_2 flood—From Continuos Injection to WAG, SPE 154181, presented at the SPE Improved Oil Recovery, Tulsa, April 14–18, 2012.

A Simple and Swift Method of Optimizing Oil and Gas Well Placement from Static Reservoir Data Utilizing Modified Well Index and Lagrange Multiplier

Steven Chandra

Abstract In this paper, a new quick look method has been generated for faster placement of oil and/or gas wells from reservoir static data only. A new customized method has been developed to easily analyze intersections of qualities that form proper characteristics of good oil and gas wells, ranging from reservoir fluid and rock properties to predictions of liquid flowrate from log data. This method is then analyzed using Lagrangian multiplier to find optimum location of production wells. The resulting model has been tested against a set of field data and has provided a better production profile from the aspects of quantity and longevity.

Keywords Production optimization · Well placement · Non-linear · Lagrange multiplier

Introduction

Oil and gas production well placement has long been an issue that presents complication in integrated reservoir optimization. This particular aspect, unlike any other design methods available to the industry, has no exact procedure or detailed analysis. This is because well placement is a very complex function of not only technical and economical but also geographical analysis. Sometimes company policies would also have to be entered into play for the decision-making process. This confusion causes some wells are not properly placed based on technical parameters and future development considerations such as artificial lift and Enhanced Oil Recovery processes.

The research in well placement optimization has begun since the start of complex development. However, first properly recorded publication on the aforementioned topic was conducted by Beckner et al. (1995). The authors used optimal annealing method to schedule well placement and drilling time for infills. The research was the developed by Pan and Horne (1998) and Bittencourt (1997) using simple hybrid

S. Chandra (✉)
Institut Teknologi Bandung, Bandung, Indonesia
e-mail: steven@tm.itb.ac.id

© Springer Nature Singapore Pte Ltd. 2018
B. M. Negash et al. (eds.), *Selected Topics on Improved Oil Recovery*,
https://doi.org/10.1007/978-981-10-8450-8_7

algorithm and genetic algorithm to properly select well placement. These methods are developed until now using more complex hybrid algorithm, such as evolutionary algorithm enhanced with artificial neural network presented in researches conducted by Ayotunde et al. (2014) and utilizing uncertainty analysis such as the work of Guyaguler et al. (2004).

The methods presented above are sophisticated, yet it requires a significant amount of data and field experiences in order to properly train the model. For cases of new field development, it is uncommon that relatively small amount of data are available, which are extracted from drill stem tests and logging data. These data are characterized as building blocks for modeling static reservoir system and sometimes do not have any indication to the productivity of the entire reservoir system. This is where the genetic algorithm model fails, as the amount of test data will not be enough to build sufficient accuracy for construction of proper well placement.

Engineers are often encountered with confusing problems while using logging data or other related measurements, such as the difficulty to find intersection between good reservoir qualities in multiple data sheets. As only a small amount of data are present, engineers need a quick look method that can properly allow them to select optimum zones for placing production without having to look for intersections of multiple data sheet.

To tackle this problem, we introduce a modified parameter that can represent reservoir static quality, based on the premises that the intersection of several static reservoir qualities can represented as multiplication of the parameters itself,

$$S\left(\emptyset, k_{xy}, k_z, P_r, HCPV, S_o, d_{GOC}, d_{WOC}\right) = \mathbb{R}, \mathbb{R} \geq 0 \tag{1}$$

If the parameters are to be multiplied, it is known that the value will be diversified in a very wide range, resulting in even harder interpretation. Therefore, this equation will be modified as an index, which will divide the function S with the maximum value of its variable, that can be easily viewed in histogram of the properties in simulation software which are readily available in the market nowadays.

$$CMOWI = \frac{\emptyset \, x \, k_{xy} \, x \, k_z \, x \, S_o \, x \, P_i x \, PV x d_{GOC} x d_{WOC} x Q_{est}}{\emptyset_{max} \, x \, k_{xy \, max} \, x \, k_{z \, max} \, x \, S_{o \, max} \, x \, P_{max} \, x \, PV_{max} x h^2 x Q_{max}} \tag{2}$$

$$CMGWI = \frac{\emptyset \, x \, k_{xy} \, x \, k_z \, x \, S_g \, x \, P_i x \, PV x d_{GOC} x d_{WOC} x Q_{est}}{\emptyset_{max} \, x \, k_{xy \, max} \, x \, k_{z \, max} \, x \, S_{g \, max} \, x \, P_{max} \, x \, PV_{max} x h^2 x Q_{max}} \tag{3}$$

The objective function itself is called Chandra Modified Oil Well Index and Chandra Modified Gas Well Index, where it will be used to determine the best placement for producing oil/gas wells. The value of the indexes itself will be ranging from 0 to 1, implying the ideal place will be closer to 1 in the index' value. The division with the maximum parameters will ease the process of calculating to enormous numbers, making it difficult to determine well placement in a quick look. These maximum values can be easily found in the reservoir simulation tools, making it easy for engineers to utilize this index.

For the indexes to work properly, constrains needed to be defined based on usual practices in oil and gas field development. The distance from contacts should be at least 30% from the total height of the productive layers, whilst maintaining at least 2 grid spacing for each well or 500 ft for oil wells and 1000 ft for gas wells. The value of initial estimated flow rate can be calculated from simple equations collected from log data derived by Crain (2009)

$$Q_o = 0.0001xk_{xy}x(P_f - P_s)/\mu \tag{4}$$

$$Q_g = 0.0001xk_{xy}x(P_f - P_s)^\wedge 2/\left(T_f + 460\right) \tag{5}$$

These constraints can also be enhanced by adding cut off values derived from accepted logging practices, such as

$$V_{sh} \leq 0.4; \ \emptyset \geq 5\%; \ S_w \leq 0.7; \ k_{xy} > 5\,md \tag{6}$$

These are normally used cut offs taken from Crain (2009), should any complications such as unconventional reservoirs or complex reservoir systems are encountered, these cut off values can be properly edited to meet the true reservoir condition.

Non-linear Optimization Utilizing Lagrange Multipliers

Lagrange multiplier is a method in mathematical optimization that is strategized to find local maxima and minima of a function subjected to equality constraints. The case of Lagrange multiplier holds true under a set of equations

$$objective\ function : f\ (x \cdot y)\ which\ is\ subjected\ to\ g\ (x, \ y) = 0. \tag{7}$$

Under the assumption that both of the objective and constraint functions possesses continuous first derivatives, it can be introduced a new variable called λ, which is a Lagrange Multiplier defined as

$$\mathcal{L}\ (x, \ y, \ \lambda) = f\ (x, \ y) - \lambda \cdot g\ (x, \ y) \tag{8}$$

where the term of λ can be added or subtracted from the function itself. If $f(x_a, y_a)$ is a maximum for the original problem, then there exists λ_0 such that (x_0, y_0, λ_0) is a stationary point for the Lagrange function.

However, due to the fact that the values of reservoir properties are not exact and diversified over a range of values, a modification has to be done to the Lagrange multiplier to satisfy the inequality constraints. This phenomenon can be easily described by using the Karush-Kuhn-Tucker conditions (KKT). The KKT condition is a first order necessary conditions for a solution in nonlinear programming to become opti-

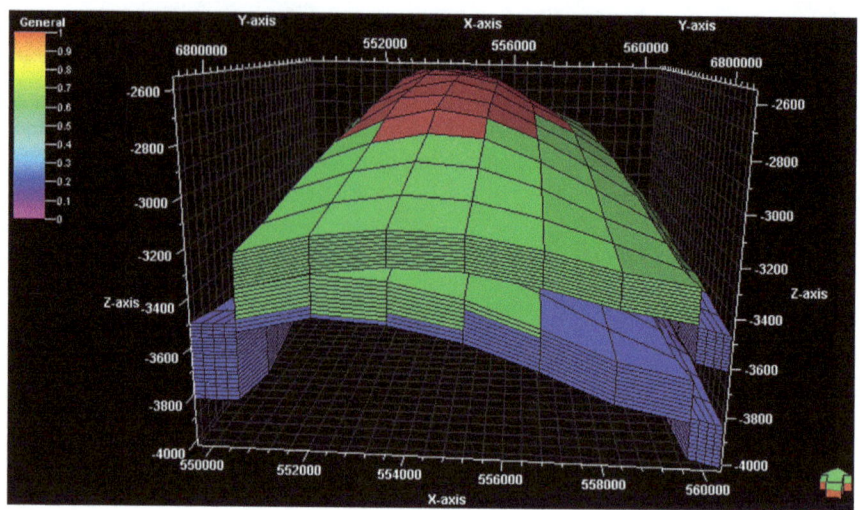

Fig. 1 Construction of X field based on fluid presence

mal, provided that some regularity conditions are satisfied by generalizing the method of Lagrange multiplier using the following derivation

$$\mathcal{L}(x, \mu, \lambda) = f(x) + \sum_{j=1}^{p} \mu_j (d_j - h_j(x)) \tag{9}$$

where d is the minimum value of the inequality constraints and h is the maximum value of the constraints. This method can be modified so that there are multiple constraints to the problem definition.

Field Data Testing

A single set of field data has been generated by commercial reservoir simulation in order to test the effectivity of the new method. The reservoir is a anticline reservoir intersected with two major faults, making it can be considered as a 3 sector reservoirs. The layering can be simplified into three layers, the first and the second are sandstone reservoirs with a single shale zone in the middle acting as an impermeable zone, as presented in Fig. 1.

The porosity and permeability will be modeled by using normal distribution function of the porosity properties with the main statistical data shown in Fig. 2.

- Mean: 0.27
- Standard of Deviation: 0.025.

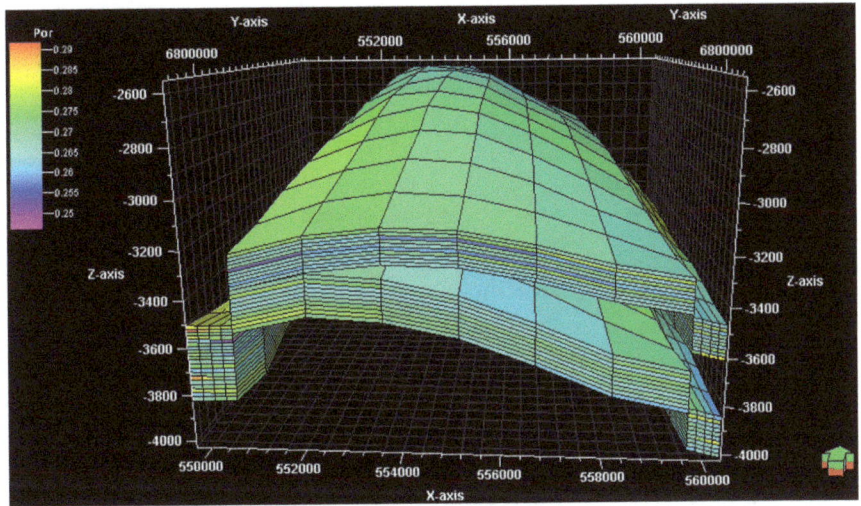

Fig. 2 Porosity distribution of field X

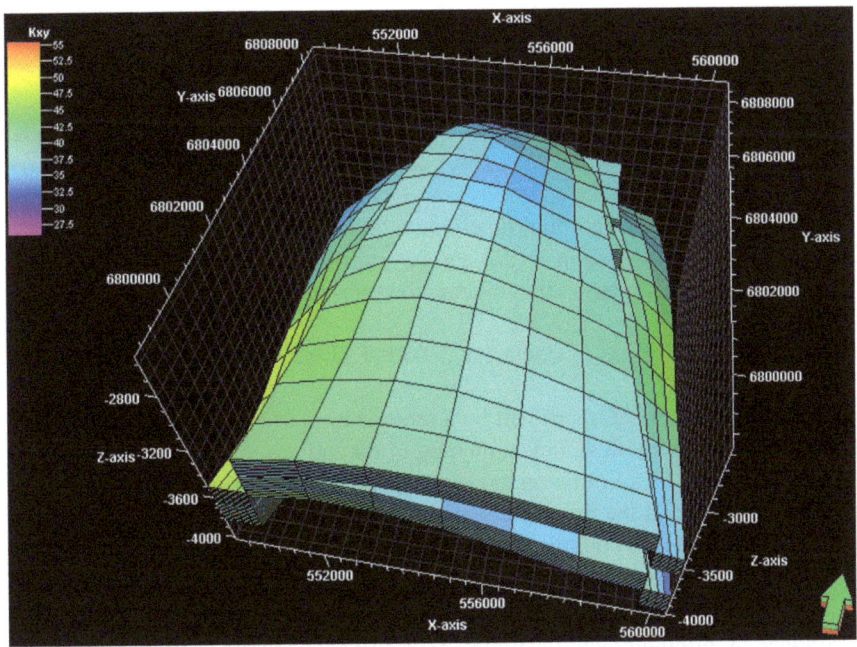

Fig. 3 Permeability distribution for field X

The permeability values, however, will be modeled using a readily available model from a reservoir study conducted in Roabiba formation as presented in Fig. 3.

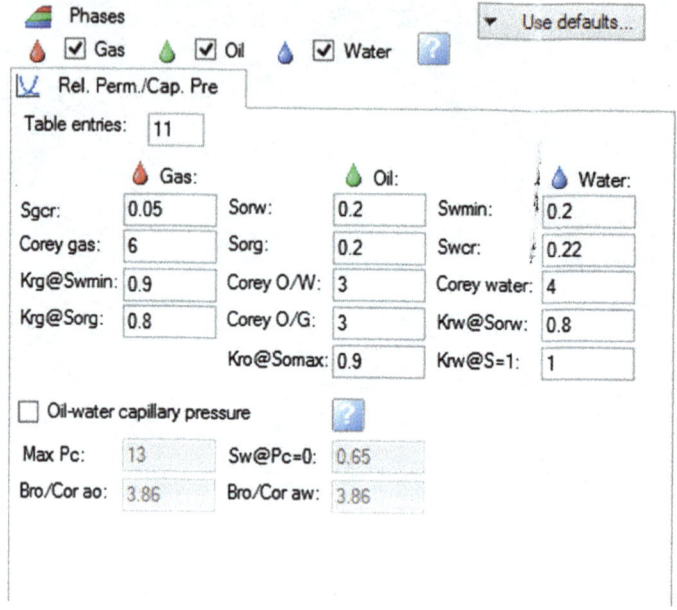

Fig. 4 Rock physics function for table X

$$k_{xy} = 268, 146.3\emptyset^{4.501} \tag{10}$$

For the reservoir rock and fluid functions, a default model generated by the simulator will be used as shown in Figs. 4 and 5.

After initialization phase, the hydrocarbon in place was determined to be (Fig. 6).

With the location of gas oil contact at 2800 ft TVD and water oil contact at 3400 ft TVD.

In developing the field, there will be a sectorization of oil and gas wells due to the fact that there are both sufficient amount of oil and gas that can be exploited in parallel. There will be a limitation, however, to the placement of perforation, since gas cap only existed in the first layer and oil zone exists in both layers. In this case, there is no possibility for commingle completion in gas zone, where in oil zone it can be mathematically performed, though in this publication we will limit the completion to a single zone for the sake of simplicity. For the sake of effective and thorough constraints, the minimum and maximum values of each parameters from the index can be summarized below

- porosity range: 0.2462–0.2915
- permeabilitas in xy axis range: 26–55 md
- permeability in z axis range: 0–29 md
- reservoir initial pressure range: 1222.85–1671.74 psi
- HCPV oil range: 0–135,056.23 cuft

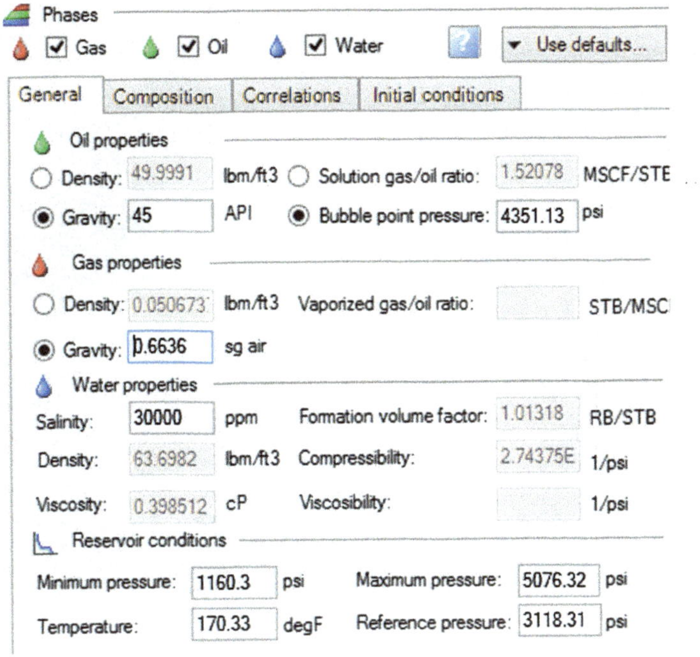

Fig. 5 Fluid properties function for field X

FIELD TOTALS
PAV = 1305.11 PSIA
PORV= 1373283356. RB
(PRESSURE IS WEIGHTED BY HYDROCARBON PORE VOLUME:
PORE VOLUMES ARE TAKEN AT REFERENCE CONDITIONS):

	OIL STB			WAT STB	GAS MSCF		
	LIQUID	VAPOUR	TOTAL	TOTAL	FREE	DISSOLVED	TOTAL
CURRENTLY IN PLACE	640010449.		640010449.	477945357.	50338508.	212454681.	262793189.
OUTFLOW THROUGH WELLS			0.	0.			0.
WELL MATERIAL BAL. ERROR			0.	0.			0.
FIELD MATERIAL BAL. ERROR			0.	0.			0.
ORIGINALLY IN PLACE	640010449.		640010449.	477945357.	50338508.	212454681.	262793189.

Fig. 6 Results of initialization phase

- HCPV gas range: 0–194,723.70 cuft.

The logging cutoffs mentioned in the previous chapter are also included in the calculation.

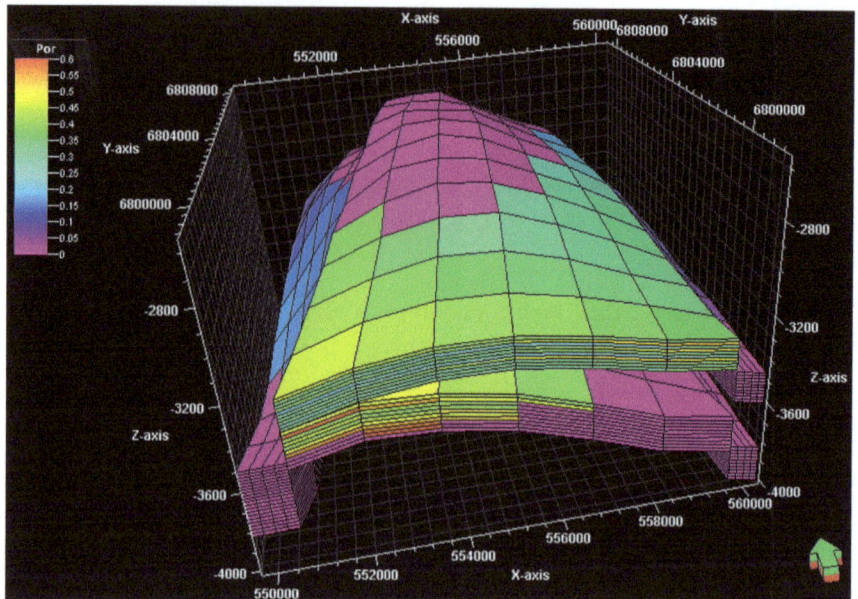

Fig. 7 Modeling for oil well index

Table 1 Detailed well location

No	Well name	Coordinate	
		X	Y
1	P1	558,705.96	6,799,005.64
2	P2	554,756.23	6,799,433.84
3	P3	552,958.09	6,798,769.96
4	P4	554,183.2	6,801,329.4
5	P5	555,419.04	6,802,161.35
6	P6	553,531.73	6,802,835.2
7	P7	554,183.2	6,801,329.4

Modeling Results

Utilizing the KKT condition and Lagrange multiplier for multiple constraints, it is found that the optimum values of each objective can be summarized as in Figs. 7 and 8.

Using only static data we propose there are 7 possible wells in the area which is (Figs. 9 and 10).

The 7 wells will be perforated based on the location of the red zone of the index itself, which is divided into well P1–4 for oil development and 5–7 for gas development (Tables 1 and 2).

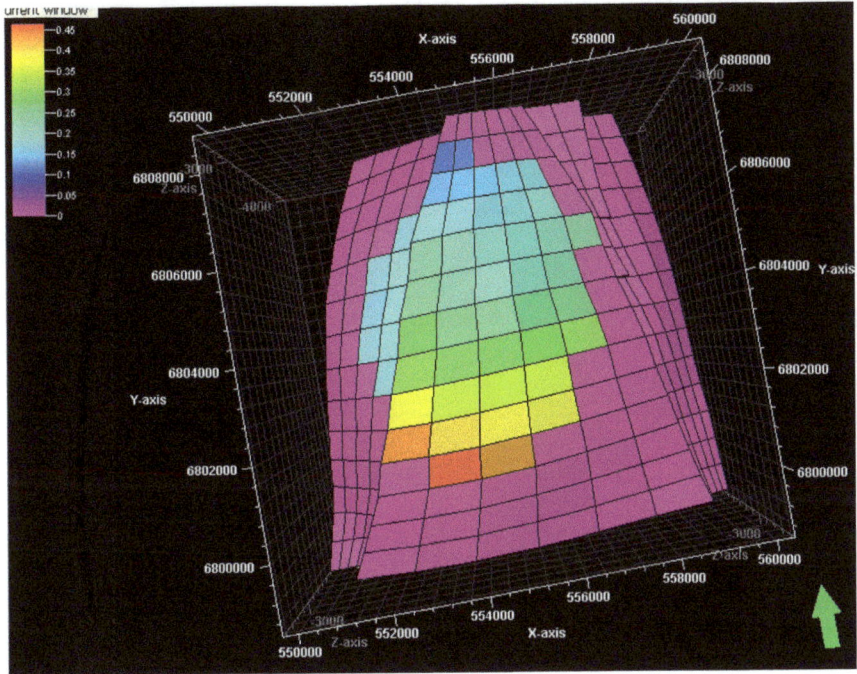

Fig. 8 Modeling for gas well index

Table 2 Well perforation location

No	Name	Perforation depth	
		Start (ft)	End (ft)
1	P1	−3212.17	−3334.86
2	P2	−3234.16	−3384.79
3	P3	−3258.73	−3422.69
4	P4	−3264.96	−3406.98
5	P5	−2705.36	−2755.36
6	P6	−2616.83	−2704.74
7	P7	−2664.21	−2794.26

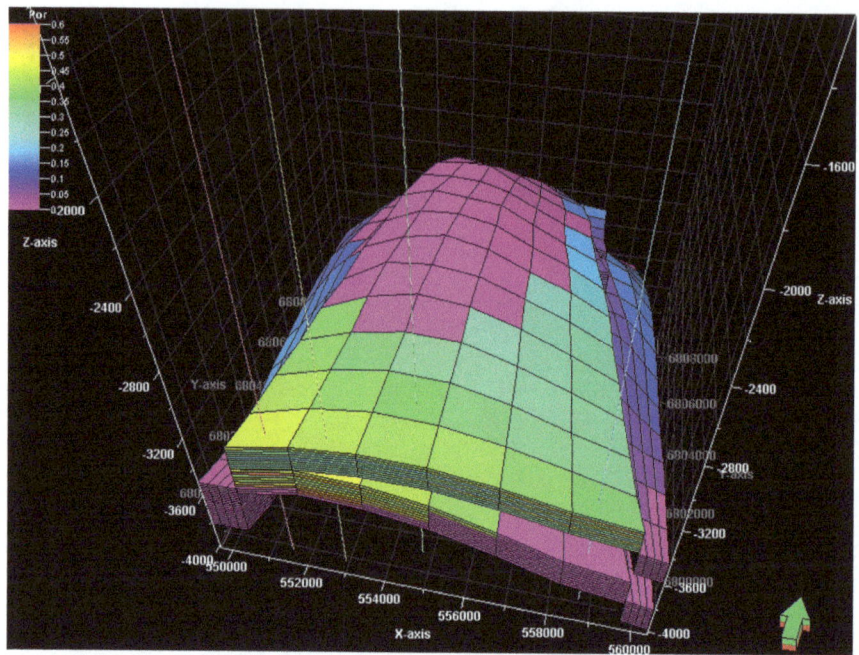

Fig. 9 Proposed oil well location

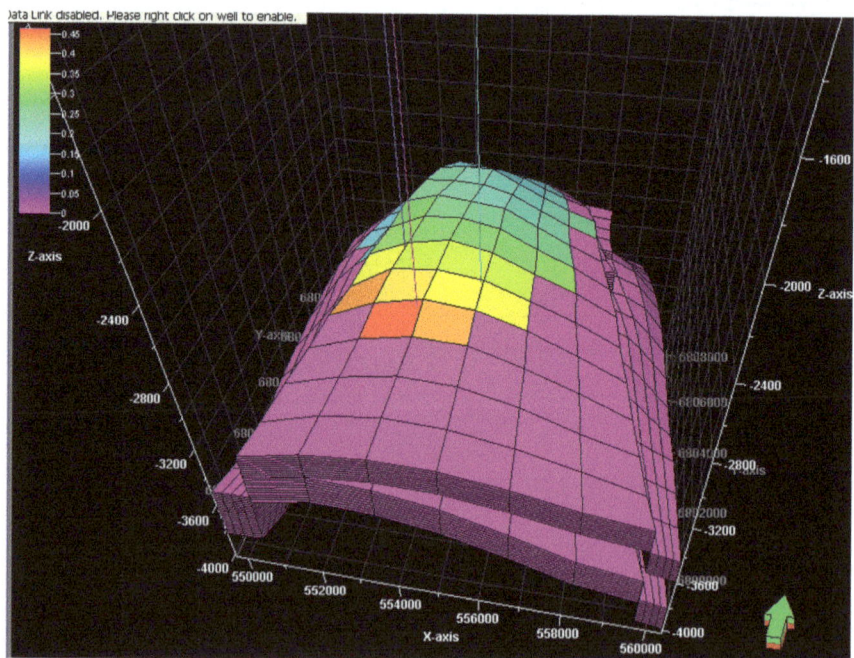

Fig. 10 Proposed gas well location

Conclusion

In this publication, a new and simple method has been developed using Lagrange multiplier and modified well index to increase efficiency in preliminary well placement. The method itself can save a lot of time and resources previously needed to inspect every single parameters of rock and fluid parameters to better predict oil and gas well performances in the future works.

Acknowledgements The author wishes to thank Dr. Amega Yasutra, ST, MT for his teaching in petroleum production engineering and Abian Adyasa Ananggadipa, S. Si, MT for fruitful discussions and new ideas.

References

Ayotunde, Ajayi, et al. 2014. *Maximizing Oil Production by Well Operating Envelope Review to Arrest Production Decline in a Deepwater Oil Field*. SPE 172409.

Beckner, B.L., et al. 1995. *Optimization of Well Placement Using Simulated Annealing-Optimal Economic Well Scheduling and Placement*. SPE 30650-MS.

Bittencourt, A., and R.N. Horne. 1997. *Reservoir Development and Design Optimization*. SPE 38895.

Crain, R.R. 2009. *Crain's Petrophysical Handbook Section RF-04 Logging Tools Theory*, accessed from https://www.spec2000.net/.

Guyaguler, Baris, et al. 2004. Uncertainty Assessment of Well Placement Optimization. *SPE Journal*.

Pan, Y., and R.N. Horne. 1998. *Improved Methods for Multivariate Optimization of Field Development Scheduling and Well Placement Design*. SPE Annual Technical Conference and Exhibition.

Optimization of Mud Injection Pressure in Oil Drilling Using Duelist Algorithm

Totok R. Biyanto, Kukuh Gharyta, Gabriella P. Dienanta, Nanda E. Tama, Arfiq I. Abdillah, Matradji, Hendra Cordova, Tita Oxa Anggrea and Sonny Irawan

Abstract Drilling fluid or commonly called mud, is used to lift drilling cutting to the surface, cool and lubricate the bit and drill string, as supporting walls in the borehole using mud cake, and control the formation pressure. The injection pressure is very important because if the injection pressure of the mud is not right, it is going to cause various effects, such as causing fracture in the wellbore, causing heat in the bit and drill string, weak wall since weak mud cake formed in the well, and being able to cause kick or stuckpipe. Because of the impact of the mud injection pressure is significant to drilling performance, the pressure of mud injection in the oil drilling needs to be optimized using Duelist Algorithms to minimize the amount of pressure drop at each hole diameter, i.e. at 17 in. hole diameter in 1269.68 ft depth and at 12.25 in. hole diameter in 2132.55 ft depth. Before being optimized, the pressure of mud injection was modeled first using bingham-plastic method for calculating pressure drop in six different sections of the wellbore, i.e. surface equipment, drill pipe, drill collar, bit, annulus around drill collar, and annulus around drill pipe. Afterwards, it was summed to calculate the total pressure drop in each hole diameter. The total initial pressure drop for 17 in. hole diameter was 978 psi and for 12.25 in. hole diameter was 1875 psi. In order to minimize the pressure drop, it was necessary to do optimization using Duelist Algorithm. The optimized variables were mud density and flow rate. Sensitivity analysis was utilized to obtain behaviour of the system. Duelist Algorithm optimization was performed using 200 iterations, 200 duelists, 80% learning probability, 10% innovate probability, 95% luck coefficient, and 10% board of champion. After being optimized using Duelist Algorithm method, for 17 in. hole diameter, mud density became 9 ppg and flow rate became 505 gpm. Therefore, the pressure drop became 695 psi. Meanwhile, for 12.25 in. hole diameter, mud density became 9.18 ppg and flow rate became 603 gpm. Therefore the pressure drop

T. R. Biyanto (✉) · K. Gharyta · G. P. Dienanta · N. E. Tama · A. I. Abdillah · Matradji ·
H. Cordova · T. O. Anggrea
Engineering Physics Department, Sepuluh Nopember Institute of Technology (ITS), Surabaya,
Indonesia
e-mail: trb@ep.its.ac.id; trbiyanto@gmail.com

S. Irawan
Department of Petroleum Engineering, Universiti Teknologi PETRONAS, Seri Iskandar, Malaysia

© Springer Nature Singapore Pte Ltd. 2018
B. M. Negash et al. (eds.), *Selected Topics on Improved Oil Recovery*,
https://doi.org/10.1007/978-981-10-8450-8_8

became 1145 psi. The results show that the optimization of mud injection operating condition in drilling process provide reduction of pressure drop in mud injection that would be give good impacts in the drilling efficiency and safety. This result will be useful for drilling engineer to set up the drilling equipment and mud properties.

Keywords Mud injection · Drilling process · Mud density and flow rate · Duelist algorithm optimization

Introduction

Nowadays, oil drilling is more concerned considering the enhancement of people's need for oil. Oil drilling itself involves various processes in determining the appropriate place for drilling. One of the important components in drilling is the drilling fluid performance. The cost of exploration to search for hydrocarbon becomes more expensive when drilling occurs in offshore with great depth and also in unsupporting environment. The cost of drilling fluid reaches fifth (15–18%) of the total cost of drilling (ADCO 2010). This drilling environment requires great fluids in its performance. Measuring the performance of drilling fluid requires evaluation of all key parameters of drilling.

Drilling fluid or mud has several functions (ADCO 2010). The first function is to lift the cutting to the surface (hole cleaning). Hole cleaning process depends on flow rate of mud. The greater the flow rate, the greater the pressure, and also the greater its friction. The second function is to cool and lubricate the bit and drill string. Bit and drill string have to be kept cold. If the pressure is not enough, it will generate heat in the bit and drill string. The third function is to act as supporting walls in the borehole with mud cake. If the pressure is too low, then the mud cake will not be formed (Ebrahim 2012), which will cause the drilling wall is not strong enough. While the last function is to control the formation pressure which depends on mud density. If mud density is too large, it will cause excessive load on the drill string that will lead to cause stuck pipe (Bailey 1993).

Because of those impacts of pressure in mud performance, it is necessary to control the amount of pressure to keep the amount of pressure drop as small as possible in order that the drilling performance can perform better. Therefore, based on the parameters that affect the mud injection pressure, drilling mud injection needs to be optimized to minimize pressure drop in order that drilling performance can be done more optimally.

Statement of Theory and Definitions

Drilling Fluids

Drilling fluid is basically a fluid (water or oil) circulated to bring cuttings out of the wellbore (Skalle 2011). They provide primary control of subsurface pressures by a combination of density and any additional pressure acting on the fluid column (annular or surface imposed) (IADC 2014). Drilling fluids are commonly called as mud.

Mud is considered as important part of drilling because it has important functions during drilling. Those functions include:

- lifting drilling cutting to the surface (hole cleaning);
- cooling and lubricating the bit and drill string;
- acting as supporting walls in the borehole using mud cake;
- controlling the formation pressure.

Mud is injected into the wellbore through mud drilling recycle that can be seen in Fig. 1.

Mud drilling recycle is started by pumping mud into the surface equipment (stand-pipe, rotary hose, and top drive). Then it gets into the wellbore (drill pipe and drill collar) and bit. After the wellbore is drilled by the bit, the cutting is taken away by the mud and carried through the annulus until they get out of the wellbore. After the cutting goes out the wellbore, it enters inside shale shaker to separate the cuttings from the mud.

In the term of mud injection pressure, there are mud properties that should be noticed; they are:

- Mud Density

The amount of force (pressure) that mud applies to the wall of the hole. Unit of mud density is pound per galon (*ppg*).

- Flow Rate

The greater the amount of mud pumped, the faster it will have to flow. The faster the flow, the greater the amount of friction. Unit of flow rate is galon per minute (*gpm*).

- Plastic Viscosity (PV)

A measure of the force required to keep the drilling fluid moving once it has started to flow. Unit of plastic viscosity is centi poises (cP).

- Yield Point (YP)

A measure of the force required to get the drilling fluid to start flowing from stationary. Unit of yield point is lb/100 ft^2.

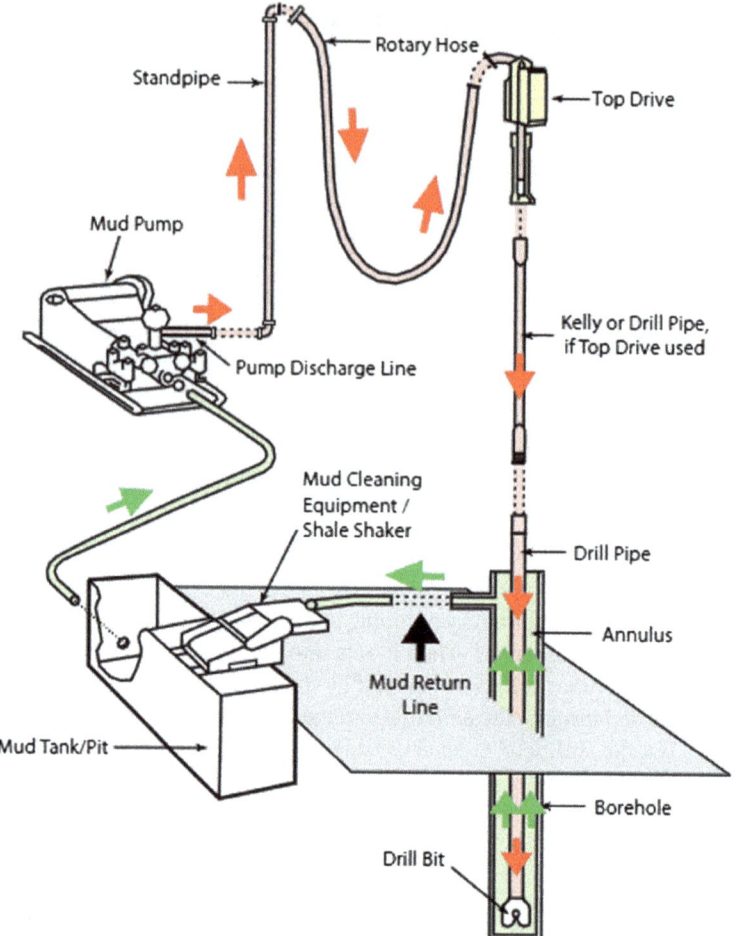

Fig. 1 Mud drilling recycle (IADC 2014)

Duelist Algorithm Optimization

Basically, optimization is an effort to achieve something better. The main goal of optimization is usually called as objective function. The objective function is achieved by changing optimized variable(s). There are a lot of optimization techniques. One of them is Duelist Algorithm.

Duelist Algorithm is an optimization method that is inspired by fighting between one or more persons. Duelist Algorithm is an algorithm inspired by how duelist improve their capabilities in a duel (Biyanto et al. 2016a, b). Duelist Algorithm is chosen due to its capabilities to find the global optimum solution surpassing the previous stochastic optimization algorithms, such as Simulated Annealing, Genetic

Algorithm, Ant Colony Optimization, Particle Swarm Optimization, and Imperialist Competitive Algorithm (Biyanto 2017). The process of Duelist Algorithm can be seen in Fig. 2. Population of duelists are registered and each duelist is encoded into binary array. The duel is set to happen between a set of duelists and will provide results that consist of champion, winners and losers. The winners and losers are determined based on their capabilities or "luck". After the duel, each winner and loser has chance to improve their skill for the next duel. The winners can improve their skills or learn new skill, while the losers can learn skills of the winners. Afterward, all duelists are re-evaluated through another post-qualification and sorted to determine the champion.

There are steps of Duelist Algorithm explained in Fig. 2. Here are explanations of each step:

a. Registration of Duelist Candidate

Each duelist in a duelist set is registered using binary array.

b. Pre-Qualification

Pre-qualification is a test that is given to keep the best duelist in the game.

c. Determination of Champion Board

Board of champion is determined to keep the best duelist in the game. Each champion should train a new duelist to be as well as itself. This new duelist will replace the champion position in the game and join the next duel.

d. Duel Schedule Between Each Duelist

The duel schedule between each duelist is set randomly. The duel duelist uses a simple logic. If duelist A's fighting capabilities plus his luck are higher than duelist B's, then duelist A is the winner while duelist B is the loser.

e. Duelist's Improvement

After the match, each duelist is categorized into champion, winner, and loser. Each category is treated separately. Losers are trained by learning from winners. While winners will improve their own capabilities by trying some thing new from the loser. And the last, each champion will train new duelist.

f. Elimination

Since there are some new duelists joining the game, there must be an elimination to keep duelists' quantity still the same as defined before. Elimination is based on each duelist's dueling capabilities. The duelist with worst dueling capabilities will be eliminated.

Fig. 2 Duelist algorithm
flowchart

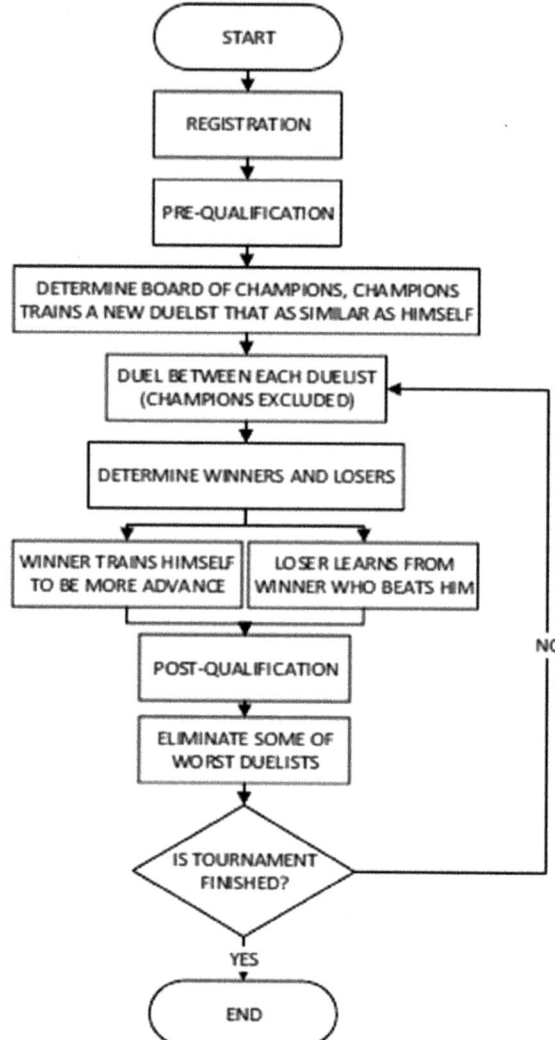

Description and Application of Equipment and Processes

The steps in doing the research can be seen in Fig. 3.

Fig. 3 Flowchart of research

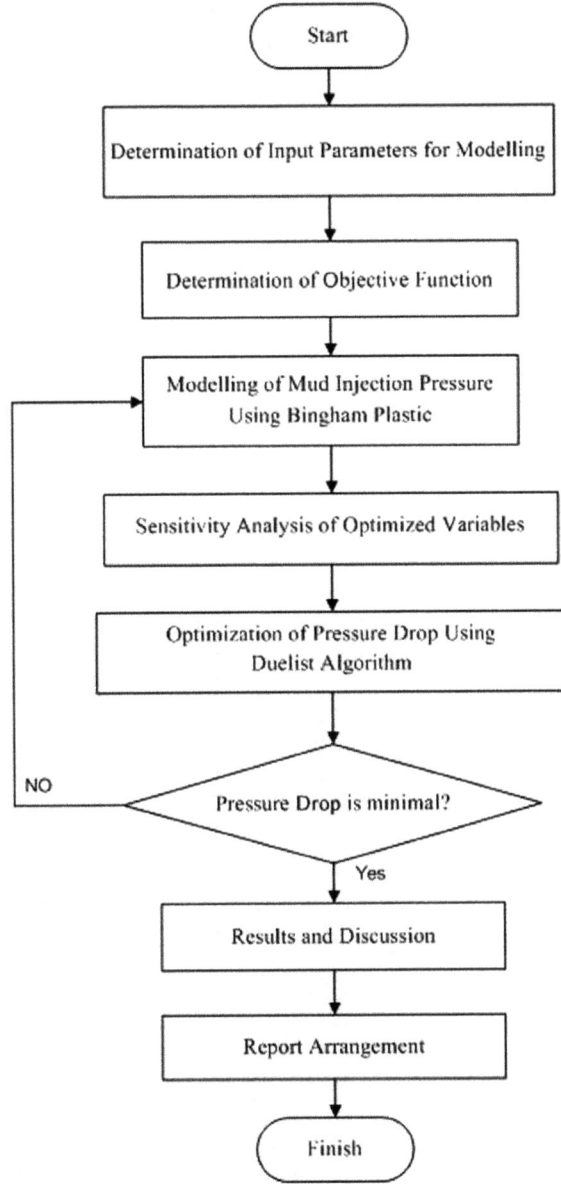

Determination of Input Parameters for Modelling

Rig is located in Java Sea that has 65 km in distance from Balikpapan city. The type of the rig is jack-up rig and vertical. The structure of the wellbore can be seen in Fig. 4.

As seen in Fig. 4, the parts that are needed to optimize are in 17 in. hole diameter at 386 m (1269.68 ft) depth and also in 12.25 in. hole diameter at 645 m (2132.55 ft) depth. The kind of drilling mud is KCl Polymer. The input parameters of wellbore to model and optimize are (Tables 1 and 2):

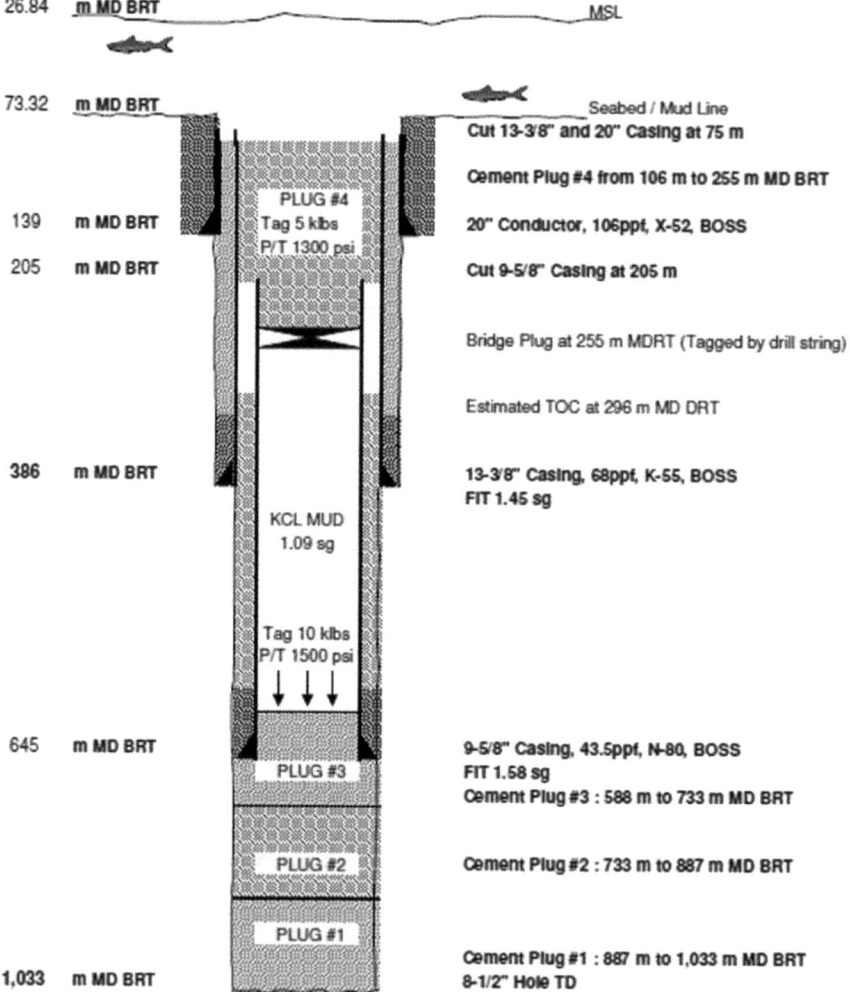

Fig. 4 Structure of wellbore

Table 1 Modelling parameters in 17 in. hole diameter

Parameter	Value	Unit
Drill pipe		
Depth	184.42	ft
Outside diameter (OD)	5	in.
Inside diameter (ID)	3	in.
Drill collar		
Depth	277.92	ft
Outside diameter (OD)	6.5	in.
Inside diameter (ID)	2.81	in.
Mud density (ρ_m)	9.26	ppg
Flow rate (q)	592	gpm
Plastic viscosity (PV)	14	cP
Yield point (YP)	20	lb/100 ft^2
Bit nozzle	3×20	1/32″
Pressure	2500	psi

Table 2 Modelling parameters in 12.25 in. hole diameter

Parameter	Value	Unit
Drill pipe		
Depth	430.45	ft.
Outside diameter (OD)	5	in.
Inside diameter (ID)	3	in.
Drill collar		
Depth	91.5	ft.
Outside diameter (OD)	8	in.
Inside diameter (ID)	2.81	in.
Mud density $\left(\rho_m\right)$	9.52	ppg
Flow rate (q)	761	gpm
Plastic viscosity (PV)	15	cP
Yield point (YP)	18	lb/100 ft^2
Bit nozzle	3×20	1/32″
Pressure	3000	psi

- 17 in. Hole Diameter
- 12.25 in. Hole Diameter.

Determination of Objective Function

Objective function is aimed to determine the goal of optimization. The goal of optimization is to make the pressure drop smaller than before by changing optimized variables in order that the performance of mud injection pressure can perform better. Pressure drop will depend on optimized variables (mud density and flow rate) based on sensitivity analysis.

Modelling of Mud Injection Pressure Using Bingham Plastic

Pressure drop is useless force because of frictions (Transocean 2009). Modelling is done to calculate the pressure drop in surface equipment, drill pipe, drill collar, bit, annulus around drill collar, and annulus around drill pipe. These followings are the equations of bingham plastic modeling (Razak 2013):

a. Pressure Drop at Surface Equipment

Surface equipment consists of standpipe, rotary hose, swivel, and kelly. The equation of pressure drop at surface equipment (ΔP_{se}) is:

$$\Delta P_{se} = E \cdot \rho_m^{0.8} \cdot q^{1.8} \mu_p^{0.2} \tag{1}$$

with:

ΔP_{se} (psi)   pressure drop at surface equipment;
ρ_m (ppg)   mud density;
q (gpm)   flow rate;
μ_p (cP)   plastic viscosity;
E   constant that depends on surface equipment that can be seen on Table 3.

b. Pressure Drop in Drill Pipe

Pressure drop in drill pipe (ΔP_{dp}) is calculated with some steps. The first step is to calculate the flow velocity with this following equation:

$$\bar{v} = \frac{q}{2448 \cdot ID^2} \tag{2}$$

Then, the critical velocity is calculated with the following equation:

Table 3 Constants of surface equipment

Type	Standpipe		Hose		Swivel, etc.		Kelly		Eq. length 3.826″ ID (ft.)	E
	ID	Length (ft.)	ID	Length (ft.)	ID	Length (ft.)	ID	Length (ft.)		
1	3″	40	2.5″	45	2″	20	2.25″	40	2600	2.5×10^{-4}
2	3.5″	40	2.5″	55	2.5″	25	3.25″	40	946	9.6×10^{-5}
3	4″	45	3″	55	2.5″	25	3.25″	40	610	5.3×10^{-5}
4	4″	45	3″	55	3″	30	4″	40	424	4.2×10^{-5}

Fig. 5 Friction factor

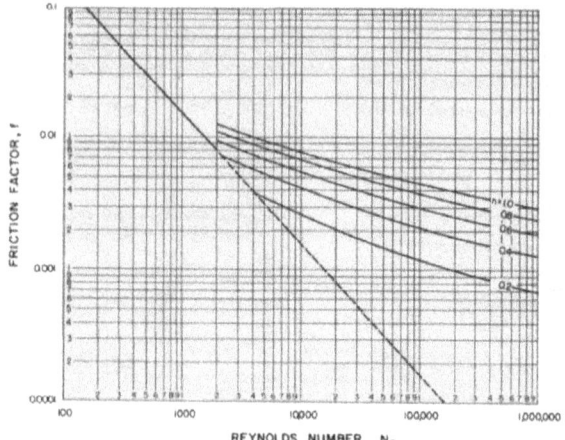

$$\bar{v}_c = \frac{1.08 \cdot \mu_p + 1.08\sqrt{\mu_p^2 + 9.3 \cdot \rho_m \cdot ID^2 \cdot YP}}{\rho_m \cdot ID} \tag{3}$$

Afterward, flow velocity and critical velocity are compared to determine the kind of the flow. If $\bar{v} < \bar{v}_c$, it means that the flow is laminar, while if $\bar{v} > \bar{v}_c$, it means that the flow is turbulent. Then the pressure drop can be calculated based on the kind of flow by these following equations:

i. Laminar Flow

$$\Delta P_{dp} = \frac{D}{300 \cdot ID}\left(YP + \frac{\mu_p \cdot \bar{v}}{5 \cdot ID}\right) \tag{4}$$

ii. Turbulent Flow

- Reynold number is calculated with this following equation:

$$N_{Re} = \frac{2970 \cdot \rho_m \cdot \bar{v}ID}{\mu_p} \tag{5}$$

Friction factor (f) is determined based on Reynold number that can be seen in Fig. 5.

- Pressure drop can be calculated using this following equation:

$$\Delta P_{dp} = \frac{f \cdot \rho_m D \cdot \bar{v}^2}{25.8 \cdot ID} \tag{6}$$

with:

ΔP_{dp} (psi)   pressure drop in drill pipe;
$\bar{v}$ (ft/s)   flow velocity;
$\bar{v}_c$ (ft/s)   critical velocity;
ID (in.)   internal diameter of drill pipe;
$YP \left(\frac{lb}{100\,ft^2} \right)$ yield point;
D(ft)   depth;
N_{Re} reynold number;
f friction factor.

c. Pressure Drop in Drill Collar

The calculation of drill collar (ΔP_{dc}) is done with the same steps as the calculation of pressure drop in drill pipe, which are using Eqs. (2)–(6). The only difference is the internal diameter. Internal diameter that is used is internal diameter of drill collar.

d. Pressure Drop at Bit

The calculation of pressure drop at bit (ΔP_{bit}) is done by calculating nozzle diameter based on the amount and size of the nozzle using this following equation:

$$d_e = \sqrt{\sum n_i \cdot d_i^2} \tag{7}$$

with d is nozzle diameter, while n is the amount of nozzle. Then, the pressure drop is calculated by using this following equation:

$$\Delta P_{bit} = \frac{q^2 \cdot \rho_m}{7430 \cdot C^2 \cdot d_e^2} \tag{8}$$

with:

ΔP_{bit} (psi) pressure drop at bit;
C discharge coefficient that is commonly worthed 0.95;
d_e (in.)   nozzle diameter.

e. Pressure Drop in Annulus Around Drill Collar

Pressure drop in annulus around drill collar (ΔP_{adc}) can be calculated by using these following steps. First step is calculating flow velocity using this equation:

$$\bar{v} = \frac{q}{2448 \cdot (OD^2 - ID^2)} \tag{9}$$

Then, diameter of annulus around drill collar is calculated with this following equation:

$$d_a = OD - ID \tag{10}$$

While critical velocity is calculated by using this following equation:

$$\bar{v}_c = \frac{1.08 \cdot \mu_p + 1.08\sqrt{\mu_p^2 + 9.3 \cdot \rho_m \cdot d_a^2 \cdot YP}}{\rho_m \cdot d_a} \tag{11}$$

Afterward, flow velocity and critical velocity are compared ro determine the kind of flow. If $\bar{v} < \bar{v}_c$, it means that the flow is laminar, while if $\bar{v} > \bar{v}_c$, it means that the flow is turbulent. Then, the pressure drop can be calculated based on the kind of flow rate.

i. Laminar Flow

$$\Delta P_{adc} = \frac{D}{300 \cdot ID}\left(YP + \frac{\mu_p \cdot \bar{v}}{5 \cdot d_a}\right) \tag{12}$$

ii. Turbulent Flow

- Reynold number can be calculated by using the following equation:

$$N_{Re} = \frac{2970 \cdot \rho_m \cdot \bar{v} \cdot d_a}{\mu_p} \tag{13}$$

- Friction factor (f) can be determined based on Fig. 4.
- Pressure drop can be calculated with this equation:

$$\Delta P_{adc} = \frac{f \cdot \rho_m D \cdot \bar{v}^2}{25.8 \cdot d_a} \tag{14}$$

with d_a is diameter of annulus around drill collar.

f. Pressure Drop in Annulus Around Drill Pipe.

The calculation of pressure drop in annulus around drill pipe (ΔP_{adp}) is done by using the same steps as the calculation of pressure drop in annulus around drill collar, which are using Eqs. (9)–(14) by firstly calculating annulus diameter around drill pipe by using Eq. (10).

After the pressure drop in each part is calculated, then the total of pressure drop can be calculated by this following equation:

$$\Delta P_{Total} = \Delta P_{se} + \Delta P_{dp} + \Delta P_{dc} + \Delta P_{bit} + \Delta P_{adc} + \Delta P_{adp} \tag{15}$$

with:

ΔP_{Total}(psi)   total pressure drop of system;
ΔP_{se} (psi)   pressure drop at surface equipment;
ΔP_{dp} (psi)   pressure drop in drill pipe;
ΔP_{dc} (psi)   pressure drop in drill collar;
ΔP_{bit} (psi)   pressure drop at bit;
ΔP_{adc} (psi)   pressure drop in annulus around drill collar;
ΔP_{adp} (psi)   pressure drop in annulus around drill pipe.

Sensitivity Analysis

Sensitivity analysis is utilized to obtain behaviour of the system. Sensitivity is anal-ized by observing the comparison graphs of each variables that affect pressure drop in six different parts of the wellbore. In this optimization, optimized variables are mud density and flow rate of the mud.

Optimization of Pressure Drop Using Duelist Algorithm

The process of Duelist Algorithm Optimization can be seen in Fig. 6.

Based on Fig. 6, Duelist Algorithm optimization is executed with 200 population of duelist. Those 200 duelists fight each other in pre-qualification process. After the battle, there are champion, winners, and losers based on their luck. Afterwards, board of champion will train new duelist and will create 10% new duelist that will join next battle. Those new duelists will fight one on one again to determine winners and losers. The winners will train themselves to be more advance. While the losers will learn from the winners who beat them. After post-qualification, 95% duelists that win previous duel will have another duel in another iteration. It will happen for 200 iterations to find the best duelist (Fig. 7).

Fig. 6 Applied duelist
algorithm optimization

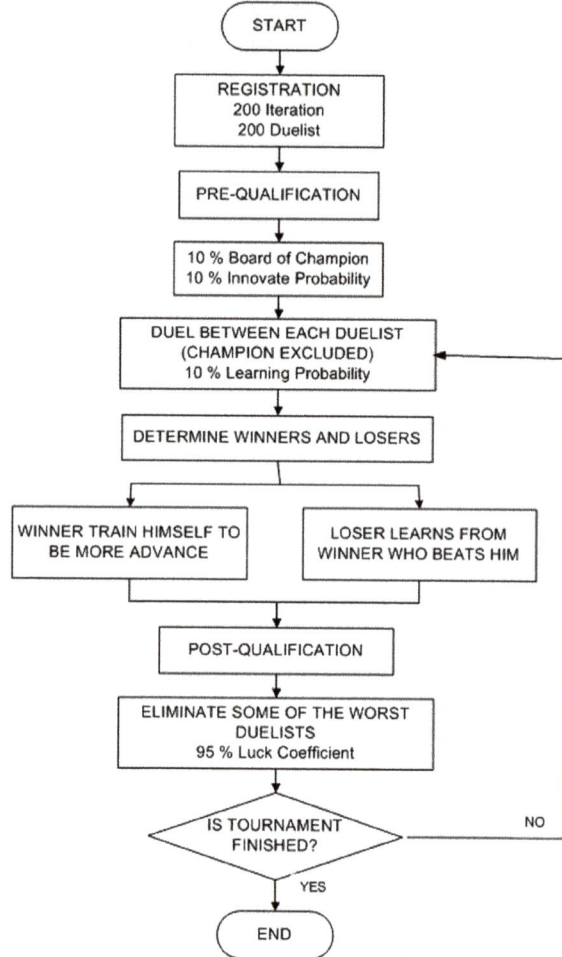

Fig. 7 Sensitivity analysis
of mud density at 17 in. hole
diameter

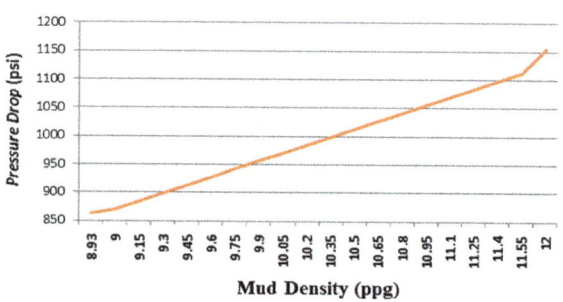

Table 4 Calculation of pressure drop in each hole diameter

Pressure drop (psi) in parts of wellbore	Hole diameter (in.)	
	17	12.25
Surface equipment	94	154
Drill pipe	105	417
Drill collar	220	125
Bit	413	702
Annulus around drill collar	25	4
Annulus around drill pipe	122	474
Total pressure drop (psi)	978	1875

Fig. 8 Sensitivity analysis of flow rate at 17 in. hole diameter

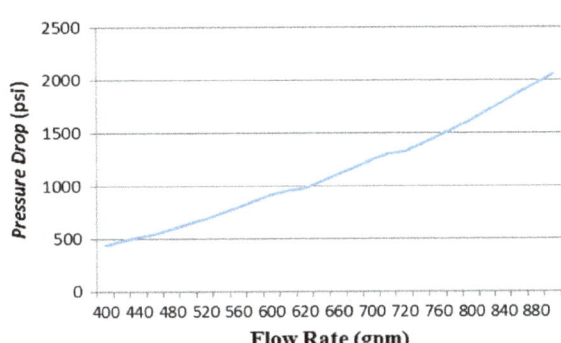

Presentation of Data and Results

Modelling of Mud Injection Pressure Using Bingham Plastic

Modelling of mud injection pressure uses Eqs. (1) until (15) to calculate total pressure drop in two different hole diameter, i.e. 17 in. hole diameter and 12.25 in. hole diameter. The input parameters can be seen in Tables 1 and 2. The result of calculation of pressure drop can be seen in Table 4.

From Table 4, it can be seen that the greatest pressure drop is at the bit, while the smallest pressure drop is at annulus. The kind of flow in all parts of drill pipe and drill collar is turbulent. The pressure drop is still too high. Therefore, in order that the performance of mud injection can be better, optimization need to be executed to decrease the pressure drop in each hole diameter (Fig. 8).

Fig. 9 Sensitivity analysis
of mud density at 12.25 in.
hole diameter

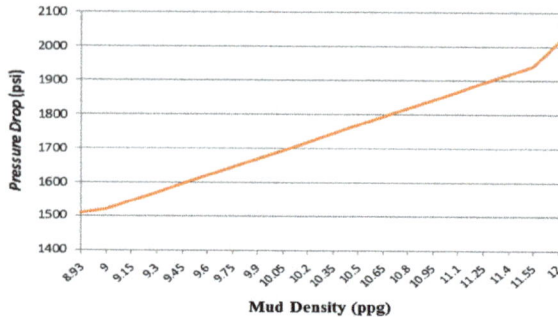

Fig. 10 Sensitivity analysis
of flow rate at 12.25 in. hole
diameter

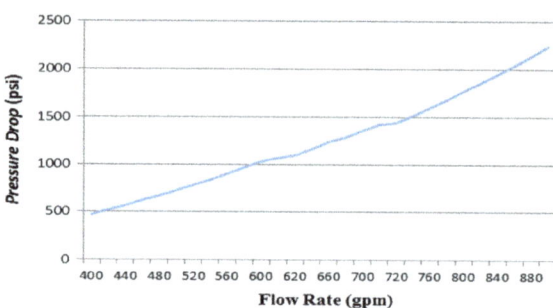

Sensitivity Analysis

Sensitivity analysis is executed to find out the effect of conversion of optimized
variables (mud density and flow rate) towards objective function (pressure drop) in
each hole diameter. Sensitivity analysis is done to calculate pressure drop in each
hole diameter by comparing it with the conversion of each optimized variables. These
are the graphs of sensitivity analysis in each hole diameter (Fig. 9):

a. 17 in. Hole Diameter
b. 12.25 in. Hole Diameter.

Considering four graphs above, it can be seen that the increasing of mud density
will give great impact towards pressure drop. It also happens on flow rate. Flow rate
has to be set properly in order that hole cleaning also can perform well (Fig. 10).

Optimization of Mud Injection Pressure Using Duelist Algorithm

Objective function of this optimization is to minimize pressure drop in order that the
drilling performance can perform more optimal by changing optimized variables.
Optimized variables are mud density and flow rate in two different hole diameters.

Table 5 Optimization result of pressure drop using duelist algorithm

Pressure drop (psi) in parts of wellbore	Depth (ft)	
	1269.68	2132.55
Surface equipment	69	98
Drill Pipe	65	222
Drill collar	129	62
Bit	292	424
Annulus around drill collar	18	2
Annulus around drill pipe	36	121
Total pressure drop	695	1145

Table 6 Optimized variables

Parameter	Value	Unit
Depth = 1269.68 ft		
Mud density	9	ppg
Flow rate	505	gpm
D = 2132.55 ft		
Mud density	9.18	ppg
Flow rate	603	gpm

Table 7 Comparison of pressure drop in each part between before and after optimization at 17 in. hole diameter

Parts of wellbore	Pressure drop (psi)	
	Before optimization	After optimization
Surface equipment	94	69
Drill pipe	105	65
Drill collar	220	129
Bit	413	292
Annulus around drill collar	25	18
Annulus around drill Pipe	122	36
Total	978	695

Constrains for each optimized variables can be seen in sensitivity analysis. After being optimized using Duelist Algorithm using software MatLab, results of optimized pressure drop can be seen in Table 5.

By changing optimized variables becomes (Table 6).

From those optimization result, there are a lot of conversions that can be seen at Tables 7 and 8.

After optimization using Duelist Algorithm, at 17 in. hole diameter, mud density becomes 9 ppg and flow rate becomes 505 gpm. Therefore, pressure drop becomes

Table 8 Comparison of pressure drop in each part between before and after optimization at 12.25 in. hole diameter

Parts of wellbore	Pressure drop (psi)	
	Before optimization	After optimization
Surface equipment	154	98
Drill pipe	417	222
Drill collar	125	62
Bit	702	424
Annulus around drill Collar	4	2
Annulus around drill pipe	474	121
Total	1875	1145

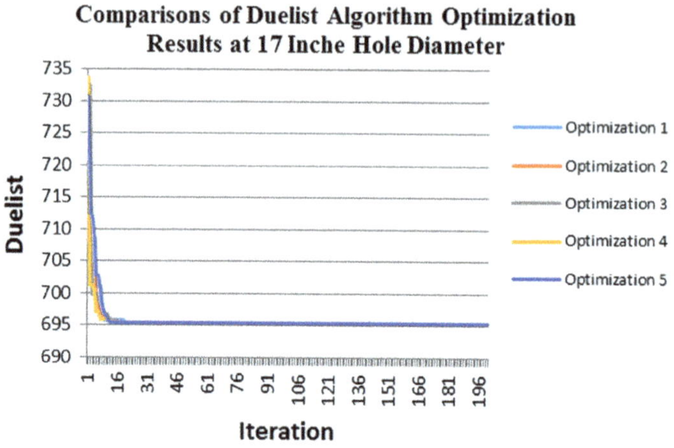

Fig. 11 Comparisons of duelist algorithm optimization at 17 in. hole diameter

695 psi. While at 12.25 in. hole diameter, mud density becomes 9.18 ppg and flow rate becomes 603 gpm. Therefore pressure drop becomes 1145 psi.

Since Duelist Algorithm is done randomly, the comparison of the results of Duelist Algorithm optimization using MatLab is done five times at each hole diameter. Comparisons of optimization results can be seen in Figs. 11 and 12.

From Figs. 11 and 12, it can be observed that those five lines of curves with different colors meet in one duelist point at the end. It shows that Duelist Algorithm optimization that processes data randomly will lead to obtain the same fitness values. Therefore, the result of Duelist Algorithm optimizations is correct.

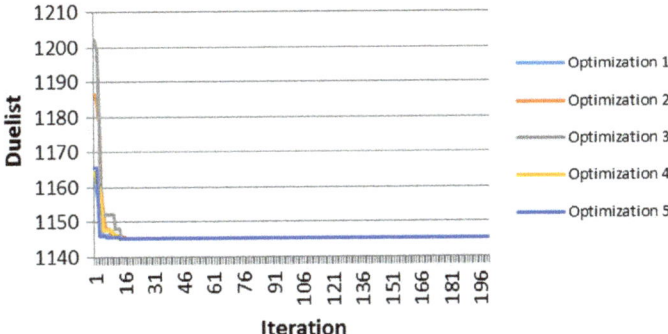

Fig. 12 Comparisons of duelist algorithm optimization at 12.25 in. hole diameter

Conclusions

The conclusion of this research is that Duelist Algorithm optimization of mud injection operating condition in drilling process provide reduction of pressure drop in mud injection that will give good impacts in the drilling efficiency and safety. This result will be useful for drilling engineer to set up the drilling equipment and mud properties.

Acknowledgements Thank you Totok Ruki Biyanto, Ph.D. and Hendra Cordova, M.T for the guide of doing this research. I also want to say thank you to Mas Arif, Mbak Rini, and Mr. Fredy who have helped me to get the data that I need for this research.

References

Abu Dhabi Company for Onshore Oil Operations (ADCO). 2010. *ADCO Drilling Manual*, 9–10.
Bailey, Louise. 1993. *Causes, Detection, and Prevention*, 1–2.
Biyanto, Totok R, et al. 2016a. *Duelist Algorithm: An Algorithm Inspired by How Duelist Improve Their Capabilities in a Duel*, 4–7.
Biyanto, T.R., H.Y. Fibrianto, and M. Ramasamy. 2016b. *Thermal and Hydraulic Impacts Consideration in Refinery Crude Preheat Train Cleaning Scheduling Using Recent Stochastic Optimization Methods*, 2.
Biyanto, Totok R, et al. 2017. *Techno Economic Optimization of Petlyuk Distillation Column Design Using Duelist Algorithm*, 2.
Ebrahim, Nuha Hussein, et al. 2012. *Optimization of Cutting Transport in Vertical, Inclined, and Horizontal Segments of The Well*, 1–5.
IADC. 2014. *IADC Drilling Manual*, 2.
Razak, Prof. Abdul. 2013. *Drilling Hydraulics*. Malaysia: Universiti Teknologi Malaysia.
Skalle, Pal. 2011. *Drilling Fluid Engineering*, 7. http://bookbon.com/.
Transocean: Singapore Training Centre. 2009. *Drilling Practices Workshop: Student Reference Material & Workbook*. Guidebook of Training PT. Switzerland: Transocean.

Optimization of Oil Production in CO_2 Enhanced Oil Recovery

Totok R. Biyanto, Arfiq I. Abdillah, Sovi A. Kurniasari, Filza A. Adelina, Matradji, Hendra Cordova, Titania N. Bethiana and Sonny Irawan

Abstract Oil production have several stage i.e. primary, secondary and tertiary. In tertiary stage, the effort to increase oil production is called as enhanced oil recovery (EOR). EOR is performed by injecting material or energy from outside reservoir. There are several EOR methods that have been developed and implemented in the oil field, including thermal recovery, chemical flooding, and solvent flooding. One of solvent flooding is CO_2 EOR by injecting CO_2 to reservoir. CO_2 EOR method has capability to increase 5–15% oil recovery. In addition, injecting CO_2 to reservoir have good impact to reduce global warming effect. However, to obtain the optimum result of CO_2 EOR needs several parameter to be optimized, such as mass flow rate, pressure and temperature injection. There are several equation that have been used to build a model of CO_2 EOR pressure drop. There are Fanning equation for injection well, Darcy equation for reservoir formation and Beggs-Brill equation for production well. The model has been validated using PIPESIM software for injection well model and have mean error 2.204%. Meanwhile reservoir formation model has been validated using COMSOL Multiphysics software and have mean error 3.863%. The optimization of CO_2 EOR using Duelist Algorithm provide increasing the net profit 42.47% from 26,548.62 USD/day to 37,826.39 USD/day.

Keywords Enhanced oil recovery · CO_2 · Duelist algorithm

T. R. Biyanto (✉) · A. I. Abdillah · S. A. Kurniasari · F. A. Adelina · Matradji · H. Cordova
Department of Engineering Physics, Faculty of Industrial Technology, ITS Surabaya, Surabaya, Indonesia
e-mail: trb@ep.its.ac.id; trbiyanto@gmail.com

T. N. Bethiana
Department of Chemical Engineering, Faculty of Industrial Technology, ITS Surabaya, Surabaya, Indonesia

S. Irawan
Department of Petroleum Engineering, Universiti Teknologi PETRONAS, Seri Iskandar, Malaysia

© Springer Nature Singapore Pte Ltd. 2018
B. M. Negash et al. (eds.), *Selected Topics on Improved Oil Recovery*,
https://doi.org/10.1007/978-981-10-8450-8_9

Introduction

Oil and gas demand increase over the time due to increase in energy consumption especially in industrial and transportation sector. Although renewable and new energy have been utilized, oil and gas are still the major energy resources to fulfill the energy consumption demand. One of method to overcome the problem is enhanced oil recovery (EOR) (Widarsono 2013).

Enhanced oil recovery (EOR) is oil recovery by injecting of material and/or energy from outside the reservoir. EOR is a way to obtain residual oil that has not been lifted through the primary method. There are several EOR methods that have been developed and implemented in the oil field, including thermal recovery, chemical flooding, and solvent flooding (Mandadige et al. 2016; Donaldson et al. 1985). Each method has their advantages and disadvantages corresponding to the reservoir and oil characteristic.

The thermal recovery mechanism reduces oil viscosity. Chemical flooding (polymer) improves volumetric sweep by mobility reduction. While the miscible gas or solvent, reduces oil viscosity, development of miscible displacement and oil swelling (reduces oil density) (Lake 1989).

Injecting of miscible gas using CO_2 has some advantages compared to other methods, this method able to increase the production of 5–15% (Lake 1989) and CO_2 as the injected gas can reach the zones that have not been reached by waterflooding and reduce the trapped oil in the rock formations. EOR using the CO_2 injection method provides a positive impact to global warming conditions. By doing the CO_2 injection into the reservoir it has reduced the amount of CO_2 in the atmosphere where CO_2 gas is a pollutant that causes the greenhouse effect (Goeritno 2000; Aprilia Dwi Handayani 2011).

CO_2 injection is obtained from Carbon Capture and Storage (CCS) Unit (Bachu 2016). The operational costs consist of CO_2 purchase costs, CO_2 injecting costs depend on pressure, and flowrate of the injected CO_2 and costs of recycling CO_2 from the oil production (Cook 2012).

In this paper, the optimization of CO_2 EOR operation condition is performed using Duelist Algorithm (DA). The optimized variables are flowrate, pressure and temperature of injected CO_2. Optimization results are expected to increase the profitability of oil production.

Method

A. Determination of operating condition range of CO_2 flood operation and reservoir formation properties

The case study used in this paper is data from Morrow County, Ohio, USA. The reservoir depth is 1067 m, reservoir thickness is 10.4 m, reservoir temperature is 87 °F, minimum miscible pressure is 1087 psia, permeability is 18.1 mD, rock formation porosity is 0.07° and 41° API oil content are the parameter from Morrow County oilfield (Fukai and Mishra 2016). The reservoir shape is assumed cylindrical and isolated with distance from injection well to production well is 100 m. The applied operating condition include injection rate of CO$_2$ is 0.5 MMscfd with injection pressure is 1071 psia and temperature injection is 31 °C. The selection of this case study corresponds to the appropriate oil field for CO$_2$-EOR, which has a deep reservoir depth, low permeability and light oil (Lake 1989).

B. Problem formulation

Problem formulation consists of objective function and constrain of optimization. The objective function of the CO$_2$ EOR is to maximize oil production as well as increase profit. The amount of oil production is proportional to the injected CO$_2$. However, more CO$_2$ injected at certain pressure incur high cost. Cost of pumping and recycling the CO$_2$ also considered in the objective function. From the data mentioned before, profit can be calculated and represented as objective function as follows:

$$\text{Profit} = [\text{Revenue}] - [\text{Cost CO}_2] - [\text{Cost Recycling}] - [\text{Cost of pumping}] \quad (1)$$

where,

$$\text{Revenue} = [\text{Oil production}] \times [\text{Oil price}] \quad (2)$$

$$\text{Cost CO}_2 = [\text{CO}_2 \text{ gas flow rate}] \times [\text{Price per unit CO}_2] \quad (3)$$

$$\text{Cost recycling} = [\text{Volume recovery}] \times [\text{Price of recycling}] \quad (4)$$

$$\text{Cost of pumping} = [\text{Pump power}] \times [\text{Time operation}] \times [\text{Electricity price}] \quad (5)$$

C. Pressure drop modeling CO$_2$ EOR using Fanning, Darcy and Beggs-Brill methods

The operating condition of CO$_2$ EOR on the inlet and outlet of the reservoir change due to some mechanism processes inside reservoir and wellbores. The CO$_2$ EOR pressure drop modeling is divided into three modelling stages: injection well, reservoir formation and production well. Pressure drop on injection well is using Fanning equation, pressure drop on reservoir formation using Darcy equation and pressure drop on production well model using Beggs-Brill equation (Srichai 2006; Banete 2014; Beggs 1973). Properties of mixture between CO$_2$ and oil are obtained from HYSYS software. That properties used in pressure drop modeling on reservoir formation and production well. The models of pressure drop are validated using PIPESIM software for injection and production well model and using COMSOL Multiphysics software for reservoir formation model.

D. Estimation of addition oil recovery of CO$_2$ EOR

Estimation of addition oil recovery of CO$_2$ EOR using Koval method. Fractional flow of CO$_2$ and oil is affected by viscosity ratio between CO$_2$ and oil. The oil production rate is calculated through additional recovery, cumulative production and mass flow rate of CO_2 EOR. The amount of original oil in place is considered in the calculation of oil production rate (Rubin and McCoy 2006).

$$N_p = \frac{\alpha + (F_i)_{BT}}{1 + \alpha} \tag{6}$$

$$(F_i)_{bt} = \sqrt{\frac{0.9}{(M + 1.1)}} \tag{7}$$

$$\alpha = \frac{1.6}{K^{0.61}} \left[\frac{F_i - (F_i)_{bt}}{1 - (F_i)_{bt}} \right]^{\left(\frac{1.28}{K^{0.26}} \right)} \tag{8}$$

$$M = \frac{\mu_o}{\mu_s} \tag{9}$$

$$K = EHG \tag{10}$$

$$E = \left[0.78 + 0.22 M^{1/4} \right]^4 \tag{11}$$

$$H = \left[\frac{V_{DP}}{(1 - V_{DP})^{0.2}} \right]^{10} \tag{12}$$

$$G = 0.565 \log \left(\frac{t_h}{t_v} \right) + 0.87 \tag{13}$$

$$\frac{t_h}{t_v} = 2.571 k_v A \frac{\Delta \rho}{q_{gross} \, \mu_s} \tag{14}$$

where:

N_p fraction of the displaceable residual oil in place recovered

$(F_i)_{bt}$ HCPV of CO_2 injected at the point at which CO_2 reaches the production wells

F_i HCPV of CO_2 injected

M Mobility ratio of the two fluids

K Koval factor

E Koval mobility factor

H Permeability heterogeneity factor

G gravity segregation factor

μ_o viscosity of the oil (kg/m s)

μ_s viscosity of CO_2 (kg/m s)

V_{DP} Dykstra-Parsons coefficient

k_v reservoir permeability in the vertical direction (m^2)

A Pattern Area (m^2)

q_{gross} gross injection rate of CO_2 (m^3/s).

E. Optimization technique

Objective function of CO_2 EOR can be obtain by determining the operating condition utilizing Duelist Algorithm (DA). The operating condition that optimized are mass flow rate, pressure and temperature of injected CO_2. The initialization for DA is

Table 1 Pressure drop parameter in injection and production well model (Dutt 2012)

Parameter	Value	Unit
Gravitation	9.8	m/s^2
Diameter of well	0.089	m
Reservoir depth	1067	m
Injection pressure	1071	psia
Mass flow rate	0.30443	kg/s
Injection temperature	31	°C
Wall thickness	0.005	m
Over-all heat transfer coefficient	2	Btu/h F ft^2

determine the initial parameters such as the number of chromosome 20 bit, population size 100, maximum generation 100, crossover probability 0.8, mutation probability 0.01 and elitism 0.95. Individual with the best fitness will be a solution to obtain the optimal objective function.

Result and Discussion

A. Pressure drop modeling in injection and production well

Pressure drop modeling in injection and production well are calculated based on parameter from Morrow County, Ohio, USA as the case study in this project. The parameters are on Table 1.

Pressure drop modeling in injection well using Fanning has been validated using PIPESIM software with mean error 2.204%. Pressure drop modeling in production well using Beggs-Brill equation also has been validated using PIPESIM with mean error 1.242%.

B. Pressure drop modeling in reservoir formation

Pressure drop modeling in reservoir formation using Darcy equation. Input pressure for this model is calculated from last segment output of injection well model. The calculation result of last segment in reservoir model becomes input for production well model. The reservoir formation properties are from Morro County, Ohio, USA on Table 2. Pressure drop modeling on the reservoir has been validated using COMSOL Multiphysics software with mean error 3.863%.

C. Calculation of additional recovery CO$_2$ EOR

Additional recovery is the increasing of oil production after CO$_2$ EOR. Based on the injection parameter before optimization, the gas flow rate is 0.5 MMscfd, then the oil production rate is 563.398 barrel per day. The crude oil price as the West

Table 2 Pressure drop parameter for reservoir formation model (Dutt 2012)

Parameter	Value	Unit
Injection-production well distance	100	m
Reservoir thickness	10.4	m
Permeability	18.1	mD
Porosity	0.07	–
Deg API	41	° API

Table 3 Calculation of net profit CO_2 injection operation

Parameter	Value	Unit
Revenue	28,482.613	USD/day
Cost of CO_2 purchase	1084.999	USD/day
Cost of CO_2 recycling	284.826	USD/day
Cost of pumping	564.165	USD/day
Net profit	26,548.622	USD/day

Fig. 1 The maximum objective function during iteration GA

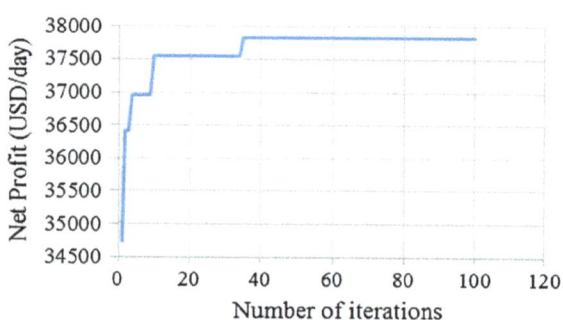

Texas Intermediate (WTI) crude oil in Septembre 2017 of 50.556 USD/barrel, so the revenue based on Eq. (2) is 28,482.613 USD/day.

The CO_2 purchase cost unit price of 2.17 USD/Mcf, recycling cost unit price of 0.505 USD/Mcf and electricity price unit price 0.0974 USD/kWh. Based on Eqs. (3–5), the CO_2 purchase cost is 1084.999 USD/day, recycling cost is 284.826 USD/day and pumping cost is 564.165 USD/day. The calculation of net profit are shown in Table 3.

D. Optimization of operating condition CO_2 EOR

The objective function of this optimization is to obtain maximum net profit. The optimized variables are mass flow rate, pressure and temperature injection. The constraint is the production well head pressure more than 100 psia. The best fitness of net profit plot from each generations are shown in Fig. 1.

Optimization result show the net profit correspond to optimized variables are shown in Table 4.

Table 4 Calculation of net profit of CO$_2$ EOR after optimization

Parameter	Value	Unit
Revenue	40,623.933	USD/hari
Cost of CO$_2$ purchase	1551.829	USD/hari
Cost of CO$_2$ recycling	406.239	USD/hari
Cost of pumping	839.477	USD/hari
Net profit	37,826.387	USD/hari

Table 5 Optimized variable after optimization

Optimized variables	Value	Unit
Mass flow rate	0.4354	kg/s
Injection pressure	1100.205	Psi
Injection temperature	35.686	C

The optimized variables that used to obtain the optimal objective function are shown in Table 5.

Conclusion

Pressure drop of CO$_2$ EOR for injection well model is using Fanning equation, Darcy equation for reservoir formation and Beggs-Brill equation for production well. Mean error of pressure model in injection well to PIPESIM software is 2.204%, the mean error of pressure model in reservoir formation to COMSOL Multiphysics software is 3.863%. The net profit at Morrow County, Ohio, USA as the case study was increased 42.47% after optimized using DA from 26,548.622 USD/day to 37,826.387 USD/day.

References

Aprilia Dwi Handayani, S. 2011. *Kendali Optimal Pada Penurunan Emisi CO$_2$ dan Efek Rumah Kaca Di Indonesia Menggunakan Metode Langsung dan Tidak Langsung.*

Bachu, S. 2016. *Identification of Oil Reservoir Suitable for CO$_2$-EOR and CO$_2$ Storage (CCUS) using reserves databases, with application to Alberta,* Canada.

Banete, O. 2014. *Towards Modeling Heat Transfer Using A Lattice Boltzmann Method For Porous Media,* Ontario.

Beggs, H.D. 1973. *A Study of Two-Phase Flow in Inclined Pipes.* SPE-AIME, pp. 616–617.

Cook, B.R. 2012. *The Economic Contribution of CO$_2$ Enhanced Oil Recovery in Wyoming's Economy.*

Donaldson, E.C., G.V. Chilingarian, and T.F. Yen. 1985. *Enhanced Oil Recovery, Fundamental and Analyses.* Netherlands: Elsevier Science Publishing Company Inc.

Dutt, A. 2012. Modified Analytical Model for Prediction of Steam Flood Performance. *Production Engineering* 2: 117–123.

Fukai, I., and S. Mishra. 2016. Economic analysis of CO_2-Enhanced Oil Recovery. *Greenhouse Green Control* 52: 357–377.

Goeritno, A. 2000. *Kemungkinan Pengenaan Pajak Terhadap Emisi CO_2 Industri.*

Lake, L.W. 1989. *Enhanced Oil Recovery.* New Jersey: Prentice-Hall Inc.

Mandadige, S.A.P., P.G. Ranjith, T.D. Rathnaweera, A.S. Ranathunga, K. Andrew, and X. Choi. 2016. *A review of CO_2-Enhanced Oil Recovery with a Simulated Sensitivity Analysis.*

Rubin, E.S., and Seat T. McCoy. A. 2006. *Model of CO_2-Flood Enhanced Oil Recovery with Application Influence on CO_2 Storage Costs.*

Srichai, S. 2006. Friction Factors For Single Phase Flow In Smooth And Rough Tubes. *Atomization and Sprays.*

Widarsono, B. 2013. *Cadangan dan Produksi Gas Bumi Nasional: Sebuah Analisis atas Potensi dan Tantangannya.*

Minimum Miscible Pressure on the CO_2 Impurities

Dinda Asmara and Riri Permata Sari

Abstract Enhanced Oil Recovery is a method to increase oil recovery from 30% till 60% depend on primary and secondary recovery. One of the proven EOR method to increase oil recovery is CO_2 injection. This injection has 2 conditions: miscible and immiscible. In this research, the purpose of MMP determination is to know the effect of temperature, pure and impurities CO_2 (methane, ethane, propane and H_2S). Simulator is used to determine of MMP. The result is the increase in temperature will increase the MMP. The influence of pure and impure from CO_2 with 80% CO_2 + 20% non CO_2 (C_1, C_2, C_3) components. With CO_2 100% as reference, the additions of 20% methane will increase 86% MMP, 20% ethane will decrease 13% MMP and for 20% propane will decrease 33% MMP.

Keywords Minimum miscible pressure · Impurities · Temperature

Introduction

Enhanced oil recovery (EOR) is implementation of various techniques to increase the amount of crude oil that can be the extraction from reservoir. EOR also called the increasing of oil recovery factor or tertiary recovery. By using EOR, oil recovery in reservoir can be extraction 30% till 60% depend on primary and secondary recovery. One of the EOR method is CO_2 injection. It is method can increase the recovery factor. CO_2 EOR has been proven in The Hansford Marmaton Field (Flanders et al. 1990).

CO_2 injection has 2 conditions: miscible and immiscible. From those conditions, recovery factor using miscible condition is greater than immiscible condition. Because it make crude oil volume is swollen, viscosity is decreased, interfacial tension is reduced, crude oil driven by solution gas, and light components are extracted to the injected CO_2 phase (Ghedan 2009). To achieve the condition, need

D. Asmara (✉) · R. P. Sari
Department of Petroleum Engineering, Faculty of Engineering, Universitas Islam Riau, Jl. Kaharuddin Nasution No. 133 Km 11, Pekanbaru, Riau 28284, Indonesia
e-mail: dindaasmara1@student.uir.ac.id

© Springer Nature Singapore Pte Ltd. 2018
B. M. Negash et al. (eds.), *Selected Topics on Improved Oil Recovery*,
https://doi.org/10.1007/978-981-10-8450-8_10

to know the Minimum Miscibility Pressure (MMP). MMP is minimum pressure for CO_2 miscibility with crude oil.

However, although the recovery factor for miscible condition is greater. Keep attention with the reservoir condition, like pressure reservoir and fracture pressure to prevent fracture or CO_2 injected can deep penetrate reservoir and not dissolve yet. Because of greater MMP value, injection pressure must be greater than MMP.

Because of that, it is important to attention the result of MMP according to reservoir condition by mixing the CO_2 that will be injected with another natural gas such as, methane, ethane and others. So, in this paper the writer try to calculate MMP by pure gas injection, and mixing the CO_2 with various component such as methane, ethane, propane, sulfur (impure), and CO_2 with flare gas by using Compositional simulator (Computer Modeling Group 2009).

Determination of Mmp

Determination of MMP has some methods, such as rising bubble technique, vanishing interfacial tension test and others. But, there are 4 primary methods that have been used in recent years to determine MMP for specific fluid displacement are: slim tube experiments, compositional simulation (Rathmell et al. 1971), mixing cell models and analytical methods.

Slimtube Experiments

Slimtube measurement is one of the standard experimental techniques that used for determining the Minimum Miscibility Pressure (MMP) of an oil and injection gas before initiation of Enhanced oil recovery (EOR) projects. Slimtube is a cylinder tube with a diameter of 0.25 in. with length ranging from 25 to 75 ft. The tube is initially saturated with the reservoir oil above is bubble point pressure. Then, the oil is then displaced by the gas injection from the tube at a fixed experimental pressure controlled by a back pressure regulator. Miscibility conditions are determined by conducting the experiment at a various pressures and recording the oil recovery. Then, MMP can be predicting with plotted curve oil recovery with pressure (Amao et al. 2012).

Mixing Cell Methods

The basic idea in this multiple mixing cell is to mix gas and oil in repeated contacts, resulting in new equilibrium compositions. This Mixing cell methods can give reliable MMP for either Condensing (enriched gas injection) or Vaporizing (lean gas injection). In case of the vaporizing drive, the intermediate component in oil is vaporized into the more mobile gas phase, and miscibility is developed when the equilibrium gas is repeatedly mixed with oil, causing the equilibrium gas composition to move toward the oil tie line. Thus, in vaporizing drive the tie line that extends through the oil control the development of miscibility. For condensing drive, the intermediate component gas is condensed into oil, and the gas tie line controls miscibility (Ahmadi and Johns 2008).

Analytical Methods

Analytical method (MOC) are based on methods of characteristic (Amao et al. 2012), or analytical method are based on the analytical solution of dispersion free 1D flow equation. MOC depend of finding the key tie lines. This key tie line are found such that key tie line when extended out of two phase region must intersect two neighboring key tie lines. As pressure is increase the key tie line are determined until one of them first intersects a critical point. MMP is the pressure at which the first key tie line become zero length (Ahmadi and Johns 2008).

Correlation

Correlations are often used to estimate MMP that the injected fluid is pure or impure. There are many correlations that can be used for calculate MMP such as Yellig and Metcalf, Helm and Josendal, Cronquist, and Glaso correlation. Glaso correlation for pure CO_2 injection is more accurate (Yuan 2004).

In this paper, the writer use Glaso correlation with if percent mol $C_{2-6} > 18\%$ with equation,

$$MMP_{Pure} = 810 - 3.404M_{C7+} + 1.700 \times 10^{-9}M_{C7+}^{3.730}e^{786.8M_{C7+}^{-1.058}} \times T \qquad (1)$$

And for $C_{2-6} < 18\%$,

$$MMP_{Pure} = 2947.9 - 3.404M_{C7+} + 1.700 \times 10^{-9}M_{C7+}^{3.730e^{786.8M_{C7+}^{-1.058}}} \times T - 121.2C_{2-6} \qquad (2)$$

where T is reservoir temperature, M_{C7+} is the molecular weight of C_{7+}.

The purpose of using correlation is to know the different MMP between simulator with glaso correlation. From Eq. (1), MMP value is 5869 psia, whereas with simulator is 5125 psia, so percent error from glaso correlation and simulator is 12.67%.

Fig. 1 Comparison of MMP values at 100% natural gas

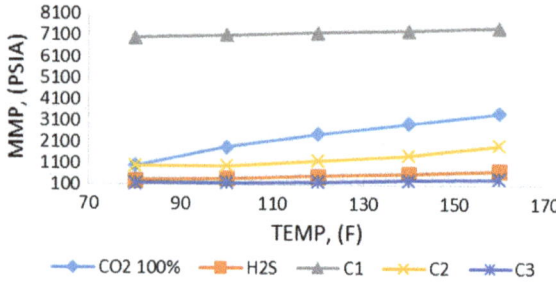

Result and Discussion

Compositional simulator is used to determine MMP. This determination is to know MMP with pure injection gas and combining. MMP is depending by temperature, oil composition, pure and impure gas injection. (Yellig and Metcalfe 1982; Alston et al. 1982; Sebastian et al. 1985) Where the oil composition (Al-Qasim et al. 2017) as the dependent variable, while the composition of gas injection and temperature as the independent variable.

In Fig. 1 can be seen the MMP value influenced by temperature. MMP for CO_2 increase with temperature and suggest that this occur, it take a higher pressure to achieve the same CO_2 density at a higher temperature. (Karimale 2010) By doing injection 100% on each component (CO_2, Methane, Ethane, H_2S, and Propane). To see how the ability on each component and know the influenced from increasing temperature with minimum miscibility pressure. So the result is the increase in temperature will increase the MMP value. Sequentially the MMP value from greater till smaller are Methane, CO_2, Ethane, H_2S, and Propane.

C_1 100% injection has a greater MMP value than CO_2 100%. Because of, the density from methane is small so, requires the big pressure to dissolve methane in fluid. It made C_1 injection can improve the oil recovery, but CO_2 injection is a very efficient to recovery method. (Holm 1982) While, component intermediate and H_2S has a smaller MMP value. However, that gas injection can't used 100% to reach target of MMP, because of being attention with reservoir condition and also about cost. To achieve the MMP value, the necessity of mixing the CO_2 with another natural gas.

So, to know the influenced of pure and impure from CO_2 is in Fig. 2. Where in the same of temperature with 80% CO_2 + 20% non CO_2 (C_1, C_2, C_3) components. With CO_2 100% as reference, the additions of 20% methane will increase 86% MMP, 20% ethane will decrease 13% MMP and for 20% propane will decrease 33% MMP.

From the result above, the effect of CO_2 impurities component on the CO_2 minimum miscible pressure at the same temperature, whereas C1 higher negative impact on the MMP, because the component increase the CO_2 MMP. But H_2S and intermediate component (C2 and C3) have positive impact on the MMP. It have been reported (Shokir and Eissa 2007).

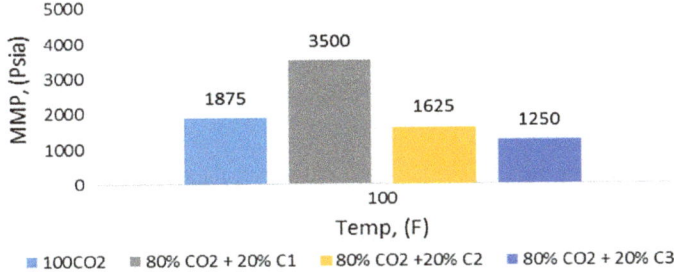

Fig. 2 Comparison of MMP 100% CO_2 versus 80% CO_2 + 20% non CO_2

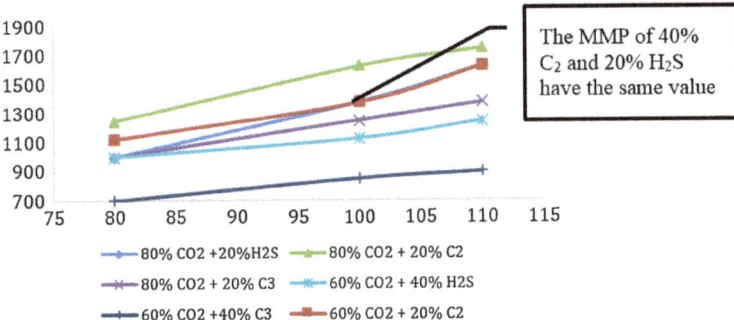

Fig. 3 Comparison of MMP with the addition intermediate component and H_2S

Fig. 4 Comparison of MMP 100% CO_2 versus Flare Gas

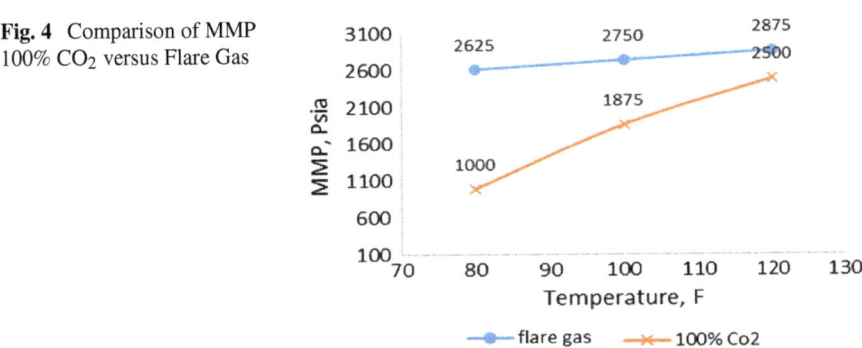

Because H_2S is a dangerous gas. So, In this research also show that the use of H_2S can be replaced by using C2 with ratio 1:2 and Will be achieve the same of MMP at the same temperature (Fig. 3).

As a comparison materials, we also present flare gas data. With the purpose to compare the MMP flare gas with MMP CO_2 100%. Can be seen in the graph below, CO_2 100% has a lower MMP than MMP flare gas. Because of the flare gas composition contains more methane (Fig. 4).

Conclusions

The main conclusion from this research are as follow:

- Temperature influence The MMP.
- The effect of mixing the CO_2 (impurities) with natural gas such as methane, ethane is helpful in achieving of MMP according to reservoir condition.
- The influence of component methane will increase MMP, While Component intermediate will decrease MMP.

Acknowledgements We would like to thank Department of Petroleum Engineering, Universitas Islam Riau and Dr. Eng. Muslim for the support in writing this paper.

References

Ahmadi, K., and R.T. Johns. 2008. *Multiple Mixing-Cell Method for MMP Calculations.* Paper SPE 116823.

Al-Qasim, Abdulaziz, and Mudhish A. Al Dewsari. 2017. *Comparison Study of Asphaltene Precipitation Models Using UTCOMP, CMG/GEM and Eclipse Simulator.* Paper SPE 185370.

Alston, R.B., G.P. Kokolis, C.F. James. 1982. CO_2 Minimum Miscibility Pressure a Correlation for Impure CO_2 Stream and Live al System. *SPEJ* 268–274.

Amao, A.M., S. Siddiqui, and H.A. Menouar. 2012. *New Look at the Minimum Miscibility Pressure (MMP) Determination from Slimtube Measurements.* Paper SPE 153383.

Computer Modeling Group. 2009. *Winprop.*

Flanders, William A., Stanbary, Wallace A., and Martinez, Manuel. 1990. CO_2 Injection Increases Hansford Marmaton Production. *Journal Petroleum Technology* 68–73.

Ghedan, S. 2009. *Global Laboratory Experience of CO_2—EOR Flooding.* Paper SPE 125581.

Holm, L.W., and V.A. Josendal. 1982. Effect of Oil Composition on Miscible Type Displacement by Carbon Dioksida. *JSPE* 87–98.

Karimale, H., and O. Terseater. 2010. *CO_2 and C_1 Gas Injection for Enhanced Oil Recovery in Fracture Reservoir.* Paper SPE 139703.

Rathmell, J.J., F.I. Stalkup, and R.C. Hassinger. 1971. *A Laboratory Investigation of Miscible Displacement by Carbon Dioxide.* Paper SPE 3483.

Sebastian, H.M., R.S. Wanger, and T.A. Renner. 1985. Correlation of Minimum Miscibility Pressure for Impure CO_2 Stream. *JPT.*

Shokir, El.-M., and M. Eissa. 2007. CO_2-Oil Minimum Miscibility Pressure Model for Impure and Pure CO_2 Stream. *OMC.*

Yellig, W.F., and R.S. Metcalfe. 1982. Determination and Prediction of CO_2 Minimum Miscibility Pressure. *Journal Petroleum Technology* 87–98.

Yuan, H., R.T. Johns, and A.M. Egwuenu. 2004. *Improved MMP Correlations for CO_2 Floods Using Analytical Gas Flooding Theory.* Paper SPE 89359.

Lightning Source UK Ltd.
Milton Keynes UK
UKHW02n1206200318
319746UK00003B/54/P